THE COMPLETE HANDBOOK OF SAND CASTING

Dedication

This book is dedicated to my wife, Hazel, who has put up with my *hoboing* over the years. Without her there would be no book.

THE COMPLETE HANDBOOK OF SAND CASTING

BY C.W. AMMEN

TAB Books
Division of McGraw-Hill, Inc.
New York San Francisco Washington, D.C. Auckland Bogotá
Caracas Lisbon London Madrid Mexico City Milan
Montreal New Delhi San Juan Singapore
Sydney Tokyo Toronto

Other TAB Books by the Author

No. 1173 *The Metalcaster's Bible*
No. 1810 *Casting Brass*
No. 1910 *Casting Aluminum*
No. 2610 *The Electroplater's Handbook*

FIRST EDITION
TENTH PRINTING

TAB Books is a division of McGraw-Hill, Inc.

Library of Congress Cataloging-in-Publication Data

Ammen, C. W.
The complete handbook of sand casting.

Includes index.
1. Molding (Founding) 2. Sand Casting. I. Title.
TS243.A47 671.2′5 78-26495
ISBN 0-8306-9841-8
ISBN 0-8306-1043-X (pbk.)

Contents

Introduction

By far sand casting is the most versatile of the various methods and techniques of forming metals which include forging, punching, rolling, stamping, extrusion and many others. Sand casting affords the designer the greatest freedom and latitude of any forming methods with an unlimited choice of metals and alloys that can be readily sand cast singly or by the millions. Sand castings are produced in a wide range of sizes from a fraction of an ounce ot over 100 tons.

EARLY HISTORY

Sand casting is a basic industry which is responsible for a great part of the industrial growth of the world. In the beginning the sand caster was also the metal smelter and refiner. At the outset the sand caster made his sand molds, then smelted virgin ore in a wide variety of home made smelters or blast furnaces and did his casting. As the demand for castings grew the foundryman had to devote more and more time and energy to the actual production of sand castings, thus he had less time to devote to the collection of ore and the smelting end of the business. This gave birth to a new business. The metal smelter who devoted his time and skills to the efficient production of pig iron, brass and bronze ingots which he sold to the ever increasing number of sand foundries. At this point the foundryman became a metal re-melter and caster as it is today. The new born smelting industry released the founder from this chore so he could devote his time to the production of castings and develop his methods to a fine point. There has been much development in sand casting methods but the basic system is the same today as it was at the outset. A sand mold is made into which the liquid metal is cast and when the metal has solidified the mold is broken up (shake out) and you have the casting.

In sand casting the developments that have taken place, and are taking place, are in improved methods of producing sand molds, mold handling equipment, automatic pouring of molds, automatic conditioning of the sand, automatic shake out and better melting equipment.

Automation and computerization have resulted in many highly efficient sand foundries. Nothing brings this into focus more clearly than a visit to a modern automotive foundry works; then a visit to a job shop foundry where men are molding by hand. Even so, it's still sand molds. You can dig a ditch with a shovel or a back hoe, it's still a ditch. In both cases you are removing dirt, the only thing changed is how you do it. Ramming up a green sand mold by hand or producing it on an automatic molding machine is still making a sand mold. Ninety-seven percent of the millions of tons of castings produced each year are sand castings with ninety percent of this tonnage being produced in green sand molds. This book is all about casting in hand made green sand molds.

GROWTH OF THE INDUSTRY

I would be amiss at this point not to say something about the cupola. The early foundry industry was restricted in its ability to cast by its melting units, which were in the beginning single batch melters and slow at that. The amount and size of castings was governed by the available melting unit. With the development of the cupola the industry began to grow in leaps and bounds. The modern cupola came about as an evolution of the forge, a product of the imagination and the genius of many founders over the years. The cupola is a miniature blast furnace.

We find reference in Genesis to Tubal-Cain as the first recorded foundry Gaffer or superintendent. It says that this old bird was an instructor in brass and iron. Cast iron and brass or wrought iron and brass? We do not know. We do know that the earliest sand casting of record is a cast iron cooking stove made in China between 200 BC and 200 AD. A Chinese stove molder of today would be hard put identifying the modern foundry product. The Chinese cupola was made of clay in the shape of a bottle. The metal in the form of native ore was melted on the fuel bed of charcoal. The blast air to produce the necessary heat was supplied by an air pump made of bamboo. The end of the bamboo air pump that was inserted into the tuyere had a clay nozzle attached. The air was generated by a feather piston attached to a smaller section of bamboo (stick). The founder built up a red hot bed of charcoal in the cupola stack, placed the ore on top, then he or his helper pumped the piston back and forth to produce the blast air. The ore melted, dripped down through the fuel and collected on the hearth. When all the metal was melted and superheated, the stack was tapped and the molten metal poured into the mold or molds.

Commercial cupolas are available in sizes from a unit that will melt one ton of iron per hour up to cupolas that will melt in excess of 50 tons per hour. Thirty years ago I built a baby cupola that had a melting rate of only 500 pounds per hour.

The basic advantage the cupola gave the foundryman was that it is a simple continuous melting unit where the metal is in direct contact with the fuel and as long as you keep charging alternate layers of fuel and metal and continue the blast air, it will continue to supply you with good molten metal. In 1952 there were more than 5,000 cupolas in daily operation in the United States. Government agencies have drastically reduced this number by imposing a mountain of regulations which has caused the rapid demise of the small and medium size jobbing foundry. A great number of foundries went to gas and oil fired reverbatory furnaces and electric melting. This increased the cost of castings. After just about blowing the cupola away as a standard melter, today there is a slow movement back to the cupola. It has been found that it was not as bad a culprit as thought.

When we talk about advancement in foundry techniques, as with any industry, things are learned and forgotten only to be rediscovered by the next generation. Some new products are not new products at all but only rediscovered old standbys, disguised as something new. Over the years I have come across many such cases.

This book will give you a well rounded working knowledge on how to sand cast ferrous and non-ferrous metals for profit or merely for your own amazement. The information in this book is gleaned from 35 years of actual foundry practice.

C. W. Ammen

Chapter 1
The Sand Foundry

The foundry is a place where the principal activity is that of making castings. To define a foundry brings to my mind many types and kinds, small, medium, large, jobbing, captive, and specialty shops. There are foundries which produce 99 percent of their castings in permanent metal molds. When I was hobo molding there were hundreds of general jobbing sand foundries which would cast anything you had a pattern for from one piece to as many as you liked with a wide weight range. The old Blackhawk Foundry in Davenport, Iowa was such a sand foundry. They would cast a fan base or a cast iron oil pan for a tractor or even an iron bust of Garibaldi. They cast gray iron, brass, bronze and aluminum, machine molding, bench molding, and what have you. Up the river in Bettendorf there was a foundry that cast only farm implement wheels. This type of foundry is fast disappearing. The captive foundry is one that does no outside casting, its entire output is consumed by the mother company. Most auto manufacturing companies have their own captive foundry. This not only assures them of a supply of castings in the quantities they desire, but gives them both quality and cost control. Foundries are further broken down into those that cast just one metal only. Aluminum, brass, iron or steel, which can be broken down still further not only by their metal speciality, but by weight ranges. Some specialize in small light weight castings, some medium weight castings and some only heavy castings. A foundry which specializes in casting steel castings from one ton and up, would probably not be interested in small light weight brass castings.

At one time you could get a job in a foundry as an apprentice. The first two years of your apprenticeship consisted of working all over the shop in every department. You worked with the molders, melters, coremakers, grinders and in the pattern shop and iron yard. Squeezing information out of any of these journeymen was tough, as each seemed to think you were after his job. Also these banty legged sand crabs all seemed to have secret methods, processes and techniques and were determined not to pass on this information to any one. To put it mildly, you as an apprentice, were a threat.

They would tell you that when you completed your time they would then and only then, pass on the ancient secrets of the art. I worked as an apprentice with an old Polish melter and when we pulled a crucible of brass from the furnace he would put in a "secret" ingredient which he kept in a snuff box and the metal would at once become clear and extremely fluid. What he added was phosphorous copper, a deoxidizing agent for copper base metals. No way would he tell me what he was doing. This was the common practice throughout the shop, some would do things or make moves which were simply a smoke screen and had no value or use whatever. Most of them, when they got to a point on a job they felt would divulge some great and rare mystery, would send you after something at the far end of the shop and when you got back the job was finished and closed up. You could only keep your eyes and ears open and glean what you could.

Once in a while you would be put to work as a helper with an old bird who would take a liking to you and he would pass on some of the mysteries. One big problem was missinformation. This was passed on to you in great quantities from all sides. Some of it was so well documented that you filed it away as great gospel and you closely guarded it from others. When you discovered a casting covered with a defect known as rat tails and buckles, you were told that this was caused by sand worms which had crawled down the sprue during the night.

Wild iron, bubbling and kicking in the ladle you were told, was caused by pollen and garlic breath being sucked into the cupola blower intake, and blown into the cupola charge causing the pigs in the pig iron to sneeze and their eyes to water so much that they ran amuck and jumped wildly around. In some cases during hayfever season, they have been known to jump out of the charging door on to the charging platform, so they said.

In one shop in Alabama the apprentice was asked to make a cast iron fry pan using a fry pan as a pattern. This he had to do on his lunch break. He would make his mold and close it, then go out into the flask yard to eat. While out in the yard some bird would take off the cope and scoop out the core and put the cope back on. When the apprentice poured the mold he would pour 25 pounds of iron into a 7 pound fry pan casting and wind up with a round blob of iron with a handle on it. He would be told that when he had completed his time the great mystery of casting hollowware would be passed on to him, then and only then. Some shops had a practice of giving every new employee, journeyman and apprentice alike, a brand new hunting knife with a leather sheath. When asked why, the gaffer would simply tell them: "You now have your hunting knife so if we catch you grinding up one of our good files and cutting up leather belting to make a hunting knife, you are fired."

No doubt they saved thousands of dollars over the years by doing this. I have seen many hunting knives made in the shop and no one knows how many company hours and material they consumed.

The practice of withholding information can be found today in most of the books I have reviewed on casting. As to whether the author is actually withholding information intentionally or simply doesn't know is hard to determine. In one book I reviewed for a friend, the author went into great detail on melting and pouring red brass. Nowhere did he mention any method of deoxidizing the metal. Not deoxidizing red brass leads to sure fire failure. Did the author leave this information out intentionally? Was he guarding a secret he felt he alone possessed? Or did he simply not know? It's anybody's guess.

The point is that the little things will throw you for a loop. The big problems you can see. What not to do seems to be the key. When you know what not to do and what won't work the rest falls in place with ease. A good founder is one who can analyze casting defects correctly and knows what corrective measures to take to rectify the problem. If you are unable to analyze a defective casting you are at a complete loss to prevent it happening again.

Well back to our apprenticeship. As I stated before the first two years was spent hodge podge, the remaining two years was spent in an elective department of your choice, molding, melting, pattern shop etc.

When you decided how you wished to finish your time, this was the direction you took for the final two years of your apprenticeship. Say you chose to go molder, you would spend the entire time left molding, bench, floor and pit. When you finished your time, four years total, much to your dismay you received your journeyman's card and your walking papers. You were told that as the name indicates you were a journeyman and must journey. You were told that you might know what went on in the shop you served your time in, and how to mold the general line of work produced in that shop, but you were far from being an all around molder, in fact you knew very little about the game; how true.

You then had to hobo mold (hit the road), a month here, a month there, north in the summer, south in the winter. It hit you like a brick just how little you knew two hours after you set foot in your first hobo job. You had to this do for a minimum of two years before you could work back in your apprentice shop. Although they were required to take you back, 90 percent of us never got back and continued to hobo for twenty years or more, some until they laid down their molding shovel and finishing trowel. This type of foundry apprenticeship has gone by the way side. Too bad. Nothing else will ever produce the all around highly skilled founder. Perhaps because of automation we will never need this type of bird again. But you cannot and never could hand a machine a loose wooden pattern and core box and say, make me one of these, please. That takes a real Honest to God sand crab.

Anyone who wishes to learn sand casting today should spend as much time as possible seeking out the few remaining unmechanized sand foundries, visit them and talk to the molders over forty years old. If you could work, even for nothing, for six months in a hand jobbing shop you would acquire more information and knowledge than any book could give. Better yet study both the book and the jobbing foundry. And, if you intend to do this you had better hurry as they are fast going down the drain.

For the most part, today's molder serves an apprenticeship on some type of automatic machine and stays there. His field of foundry knowledge begins and ends right there at his work station. When he finishes his apprenticeship, if you can call it that, he receives a journeyman's card as a molder.

Years ago, Sam Pitre and I had a jobbing foundry in New Orleans, Louisiana. We hired a molder from Ohio who had spent 20 years making a 6 inch cast tooth sprocket on a squeezer machine. He was a lost ball in high weeds in our shop, but if we ever got a job to make 6 inch sprockets we would have had it made. When this bird realized he was really not a molder, but a 6 inch sprocket maker, he asked us to put him on as an apprentice and in time he made a damn good all around jobbing molder. In the seven years he was with us he could make everything from a 2 inch rail ball to a 6 foot boat propeller.

Chapter 2
Molding Sand

When we speak of sand casting the first thing that comes to the minds of most people is the sand on the beach or the desert. There are many other places to find sand.

You can prospect for your own molding sands by looking for deposits along creek beds, river banks, cliff strata well as the beach and desert.

Sometimes you can find a nice seam of beautiful molding sand directly underneath the overburden by digging around. I always stop when there is an excavation of any size and take samples when they look promising. I watched a parking lot in Denver being built. The trucks were dumping fill in the area before black topping. It turned out that this fill was some of the nicest natural bonded molding sand for brass and aluminum casting ever. I asked where the sand was coming from and was told a bluff fifty miles west of Denver.

The sand seems to be where you find it. What you are looking for is a fine silica sand which is running about eight to twelve percent clay, actually a silty loam, free from rocks and boulders with a minimum of decayed vegetation and roots. As with any type of prospecting regardless of what you are looking for, you must have a knowledge or relationship with it, and the areas and conditions under which it is most likely to be found. Then you will go out and find it in the most unlikely places.

It's also fun and gets you out into the open places with fresh air, if you can take it. I think the entire state of Alabama is covered with good molding sand of all grades.

PROPERTIES OF MOLDING SANDS

To cast sand we make a mold around a pattern, open the mold, remove the pattern, close the mold and fill the cavity left in the sand with molten metal. When the metal has solidified we shake out the mold and have an exact duplicate in metal of our pattern. Anyone who has built sand castles on the beach can attest to how fragile they are. But if we took our beach sand and mixed enough clay with it to give each grain a coating of clay, which when damp is sticky, we would soon realize that we could make great sand castles, not nearly as fragile as our beach castles, which depended upon water alone as a medium to bond the grains together.

With additional experimentation we would find that a mold made of our beach sand, clay and water, could be used to hold molten metal. The next question that would come to mind is why when the mold is filled with hot metal it doesn't crumble or explode because of the moisture content. Very simple, this is what happens: as the molten metal enters the mold cavity the radiant heat from the metal dries the mold material in advance of the metal flow. The moisture is changed to steam and moves out of the mold trough the mold walls because of the porous nature of the molding sand.

Permeability

The ability of the mold material to allow the steam to pass through the walls is called permeability. Permeability can be measured with a meter which measures the volume of air that will pass through a test specimen per minute under a standard pressure.

Some instruments are designed to measure a pressure differential which is indicated on a water tube gauge expressed in permeability units.

In this book we are for the most part concerned with natural bonded sands, to be used in green sand molding. A natural bonded sand contains enough bonding material that it can be used for molding purposes just as it is found in the ground.

Natural molding sands contain from eight to twenty percent natural clay, the remaining material consists of a refractory aggregate, usually silica grains.

Any natural sand containing less than 5 percent natural clay is calld a bank sand and is used for cores or as a base for synthetic molding sand.

Commercial molding sands mined by various companies usually acquire the name of the area where they are mined. The most popular natural bonded molding sand is called Albany, and is mined in several different grades by the Albany Sand & Supply Co., Albany, N.Y. The origin of this sand is from the pleistocene ice sheet of approximately twenty thousand years ago which swept down from the north and completely overran what is now known as the Albany District. The result after eons is a seam of fine molding sand approximately 15 inches thick directly under an overburden of 8 inches of top soil.

Before discussing the prospecting of molding sand, let's look at a few sand characteristics necessary for the production of castings of various metals and sizes.

Light gray iron

Fineness	175
Clay	12%
Moisture	7.4%
Permeability	15
Green compression	4.0

Medium gray iron

Fineness	111
Clay	15%
Moisture	7.5%
Permeability	40
Green compression	4.0

Heavy gray iron

Fineness	73
Clay	18%
Moisture	7.6%
Permeability	70
Green compression	5.0

Heavy brass

Fineness	108
Clay	12%
Moisture	7%
Permeability	51
Green compression	4.0

Light to medium brass

Fineness	218
Clay	13%
Moisture	8%
Permeability	18
Green compression	4.0

Aluminum

Fineness	232
Clay	19%
Moisture	8%
Green compression	5.0

After studying the above closely you will notice two differences between the sands; the grain fineness and the permeability required in a sand used to make a mold for a gray iron casting compared to the fineness and permeability required in a sand for use in making a mold for aluminum.

Fineness

This is a measure of the actual grain sizes of a sand mixture. It is made by passing a standard sample, usually 100 grams, through a series of graded sieves. About ten different sieve sizes are used. As most sands are composed of a mixture of various size grains there is a distribution of sands remaining on the measuring sieves.

The fineness number assigned to a sample is approximately the sieve (screen) which would just pass the sample if its grains were all the same size.

When you fully understand the relationship between the required permeability for a given metal and it's pouring temperature and the relationship of grain fineness to permeability, you will be able to establish what sand you need for a given metal and casting size. As an example gray iron is poured at a temperature of 2700 degrees Fahrenheit and is three times the weight of aluminum which is poured at 1400 degrees Fahrenheit.

Although we are primarily interested in natural bonded sands, their use and properties, we will, however, cover some aspects of synthetic molding sands.

Both sands have their good and bad features and the choice depends upon the class of work and the equipment available.

The basic components of most molding sands are silica and a clay bond. However molding sands can also be made up of other types of refractory materials such as zircon, olivine, carbon, magnesite, sillimanite, ceramic dolomite and others.

Molding sand is defined as a mixture of sand or gravel with a suitable clay bond. Natural sands are sands found in nature which can be used for producing molds as they are found. Synthetic molding sands are weak or clay free sands to which suitable clay or clays are added to give them the properties needed.

You should be aware, however, that in the past twenty years there has been a real crop of sand medicine men who sell all types of additives and dopes for molding sands to cure the foundryman's problems.

Most of these products are simple additives which have been disguised in some manner. Make it a point never to buy any sand additive or product if the manufacturer refuses to divulge exactly what the product is. Case in point: Foundryman has a problem with *buckles* (casting defect). If he was on his toes he would know how to cure this problem. In comes the sand witch doctor with a product called "Zoro No Buckle," (fictitious name of course) bright red in color and you are told that after years of research his firm has come up with this special stuff which will chase away your scabs, not only right away, but that continued use of Zoro in the prescribed amounts will keep you out of trouble. Now here's the rub, it will do exactly what is claimed and the scabs go away and you are sold on Zoro and its use. Should you have the audacity to ask the salesman (witch doctor) what its made of, he will be visibly shook and tell you that its a secret formula, etc. etc. So, you are happy with the product's results so why press the issue. Who is going to go through the trouble to have Zoro analyzed? What the hell, it is only x dollars a sack and does a fine job. Now what did you really buy and add to your sand? Wood flour, which has been dyed red and named Zoro No Buckle. Your foundry supply house would have sold you wood flour for one half the price of Zoro which he probably carries also, had you simply asked for wood flour. We are living in an age of rediscovery and many of these rediscoveries are disguised to look like new products. In order to know and control your sand you must know exactly what you are adding and using. There must be no unknowns, otherwise you are in the dark. The rule applies to the entire foundry operation. There are several hundred products floating around which would fit into the Zoro's classification.

Do not buy any foundry product if the manufacturer will not divulge the contents. This information is vital to the control and understanding of your operations.

Avoid complicated sand mixtures, they only lead to confusion and are most difficult to control. A simple sand with the proper grain size and distribution with sufficient bond and moisture, will give much better results than one which is complex. Complex sands are usually a product of experimentation with various additives trying to accomplish some particular illusive result. This is usually due to insufficient understanding of molding sands, their formulation, limitations and uses.

Keep the types, kinds, number and amounts of sand additives to a minimum for best results.

Avoid the use of products which are sold as cure alls. No product can offset poor practice.

I took over a sizeable bronze shop as superintendent many years ago that. The scrap was running 15 percent and what they called good, looked bad. The sand system was so loaded with unknown cure alls that it was no longer molding sand but mud. It contained silica, iron oxide, seacoal, silt, wood flour, dead clay, organic matter, rocks, cat droppings, flour, oil, and several complete unknowns. I sent the whole works, all 300 tons to the dump, and replaced it with 170 mesh washed and dried sharp silica sand which I bonded with 4 percent southern bentonite and 4 percent moisture.

The scrap fell to 1½ percent, the quality was up 100 percent and 90 percent of the 1½ percent scrap was due to causes other than sand. I'm smart? No, just keeping it simple and basic. Sand, bond and moisture easy to control.

Refractoriness

This is the ability of sand to withstand high temperaures without fusing or breaking down.

From this we can deduce that a sand used for casting steel must be more refractory than one for brass or aluminum because of the greater pouring temperature involved. Also, a sand used to cast large heavy castings must be more refractory than one one used for light thin castings of the same metal.

As we are primarily dealing with natural bonded sands the refractoriness of the sands can vary over a wide range. When a naturally bonded sand contains appreciable amounts of fluxing agents (various mineral salts, organic material and oxides) that lower the fusion point of the sand, it may melt or fuse to the casting. There are various costly instruments used to determine the refractoriness of molding sands. In the absence of such testing equipment we must

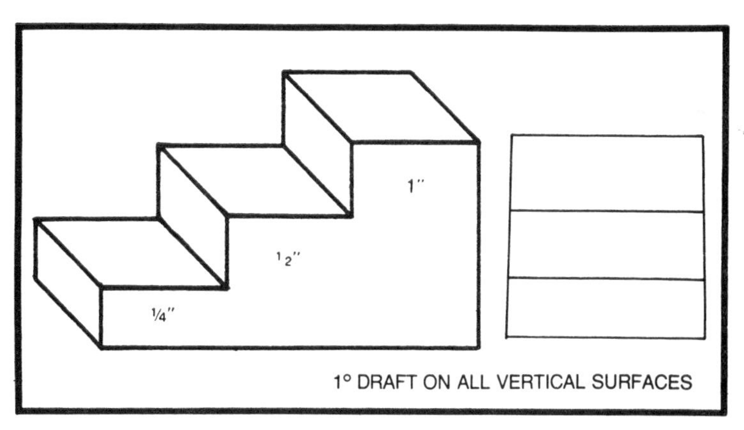

Fig. 2-1. Step pattern.

pour samples of various thickness into the sand in question and examine the surface of these test castings for sand fused to the casting. This actual experience will be more useful to you than all the instruments in the world. This is what we call getting your hands in the sand. In order to start our testing we need a wood pattern. See Fig. 2-1. This is a step pattern 6 inches long with three different step thicknesses.

We make a mold of our step pattern in the conventional manner with the sand in question and cast it in brass or bronze at 2200 degrees Fahrenheit, allow the casting to cool to room temperature in the mold, then shake it out. See Fig. 2-2.

The cold casting is carefully examined for adhered sand which will not peel off easily with a wire hand brush. Any area on which the sand will not peel off is examined carefully with a magnifying glass. This examination will readily show if the sand has indeed melted or vitrified under the heat of casting and has in fact welded or fluxed itself to the casting.

Let's say that the ¼ inch thick section and the ½ inch thick section peeled nicely and is free of any adhered sand and presents a nice smooth metal surface, but the 1 inch section does not peel clean and shows vitrified sand on its surface. This would indicate that the sand has a suitable refractoriness for light brass and bronze up to ½ inch in section thickness but is unsuitable for 1 inch and thicker sections.

Should all surfaces peel clean then you must work up a new and thicker pattern. By experimenting you can readily determine the limits of the sand in question.

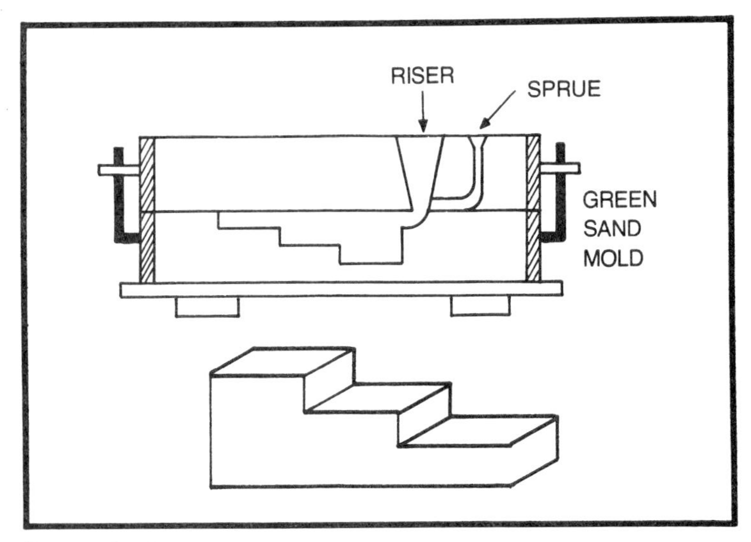

Fig. 2-2. Casting step pattern.

The reason brass or bronze is used for the test and not aluminum, is because aluminum pours at such a low temperature it would be hard to find any sand that would fuse at these lower temperatures. In making the refractoriness test the mold must be carefully made and rammed properly. A soft mold (under rammed) or a mold which has not been rammed evenly could give you a defect called penetration which you might falsely identify as poor refractoriness. In this case the surface is rough and contains metal mixed in the sand. But enough of this, we will get into it deeper as we go on.

Some foundries have and can properly use sintering apparatus, a device to determine refractoriness of sands; they may all have one today but in my 35 years of hobo molding I only served time in two shops which had such an animal. In both cases it was not in use nor had it been in use for some time, judging from the dust cover. At any rate it is a post mortum device for the most part. If the refractoriness of sand is too low for the type and weight of metal being poured, it is quite evident from the casting results. Such sand must be replaced with one that works. I am not against such devices for research work and in some types of operations they could be invaluable but nothing beats pouring some test wedges.

Green Bond Strength

This is the strength of a tempered sand expressed by its ability to hold a mold in shape. Sand molds are subjected to compressive, tensile, shearing, and transverse stresses. Which of these stresses is more important to the sands molding properties is a point of controversy.

This green compressive strength test is the most used test in the foundry.

This is how it is done. A rammed specimen of tempered molding sand is produced that is 2 inches in diameter and 2 inches in height. The rammed sample is then subjected to a load which is gradually increased until the sample breaks. The point where the sample breaks is taken as the green compression strength.

The devices made to crush the specimen are of several types and quite costly. The readings are in pounds per square inch. Now it sure wouldn't take much thinking or doing to come up with a home made rammer or a compression device. As you are only interested in whether the sand is weak, strong, or very strong, the figures you give for values are only relative and you can call them whatever you like. If a sand is suitable in green strength for your purpose, it doesn't make much difference whether the figures are in pounds per square inch or coconuts.

Tensile Strength

Tensile strength is the force that holds the sand up in the cope. And, as molding sands are many times stronger in compression than tensile strength, we must take the tensile strength into account. Mold failure is more apt to occur under tensile forces.

Where compression strength is measured in pounds per square inch, the tensile strength of molding sands is measured in ounces per square inch.

The tensile strength which is the force required to pull the sample apart, is determined very easily.

This is the old time molders test. When you get good with this one you will find that it is the one you will use the most.

Pick up a handful of riddled tempered sand and squeeze it tightly with your palm up, open your hand and observe if the sand took a good sharp impression. See Fig. 2-3.

Now grab the squeezed sample with both hands between the thumb and first finger of both hands and pull it aprat. See Fig. 2-4.

Now examine the break, it should be sharp and clean, not crumbly. By observing the break and noting the force required to pull the sample apart you can tell a lot about the tensile strength and general condition of the sand. Make another sample, this time grab it

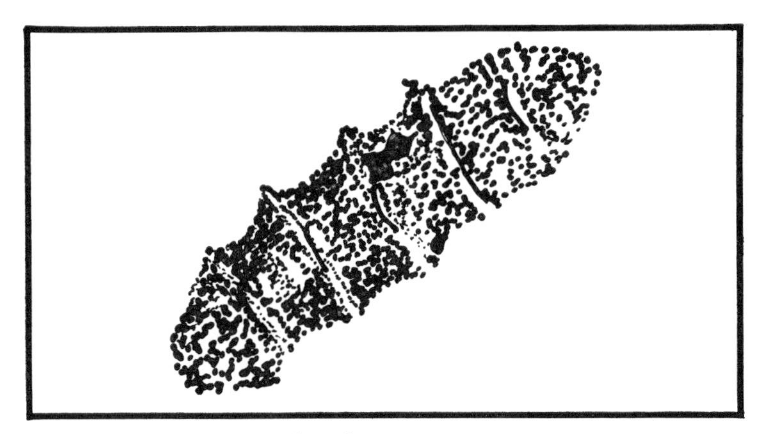

Fig. 2-3. Squeezed sample of sand.

about midway with the right hand thumb and first finger, shake it until the end that protrudes breaks off. Note how much force was required to make it break. See Fig. 2-5.

With this procedure the molder has performed a green compression test, a tensile, and a shear test, with his hand, evaluating the sand in his mind. By virtue of his experience with molding sand it is uncanny how a good molder can choose a good sand for the job at hand. Years ago we used to watch a new bird or a visiting molder when he came into the shop. If he was the real McCoy, before he would even take the job or say hello he would walk over to the nearest pile of conditioned green sand and shove his hand into the pile wrist deep, grab a handful, squeeze it, and pull it apart. Then if he was satisfied with the green strength, he would place one of the broken halves to his mouth and blow, checking the permeability.

He would, if satisfied, take the job, if not he would tell you in no uncertain terms what he thought of the sand you called green sand and slowly walk away. If he were a visiting molder or sand crab he would go through the same ritual. If satisfied he would be off on some other subject. If not, he would want to know how you made molds with this junk and then take a bite of plug and go off on a long winded spiel on what's wrong with your sand and what you should do about it. Plus he would, if you lent him your big hairy ears, tell you about some of the finest green sand this side of the horn that he had the pleasure of molding with at the old Guggenheimer Iron Works up in Wala Wala, Washington, back in the good old days where they poured nothing under six tons in weight at 2875 degrees Fahrenheit and all the castings peeled as smooth as the belly of a she mouse, bla bla.

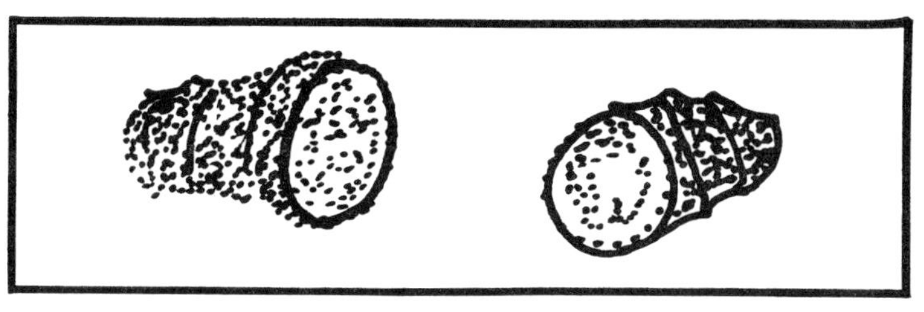

Fig. 2-4. Sharp clean break.

And if this bird came into the shop 600 times a week he would make 600 sand tests.

They were never content to squeeze test your floor sand and just shake their head, no way; you got the complete treatment each time.

At the old Blackhawk Foundry we had a kid who took samples from each heap three times a day and took them into the sand lab for testing. The results of these tests were delivered the following morning after we had poured 250 tons of metal. I never did determine whether we were to wait for the results and pour every other day or what.

At any rate the results of these tests were carefully filed away for safe keeping and took up the better part of nine filing cabinets. But as we all took our own tests (by hand) and told Tony on the sand mill what we thought was wrong or right, we managed to make some good work despite the sand lab. Don't get me wrong, in the modern high production automatic foundries of today it is necessary to have a good lab man.

Now for a little more general information in regard to molding sands for you to digest.

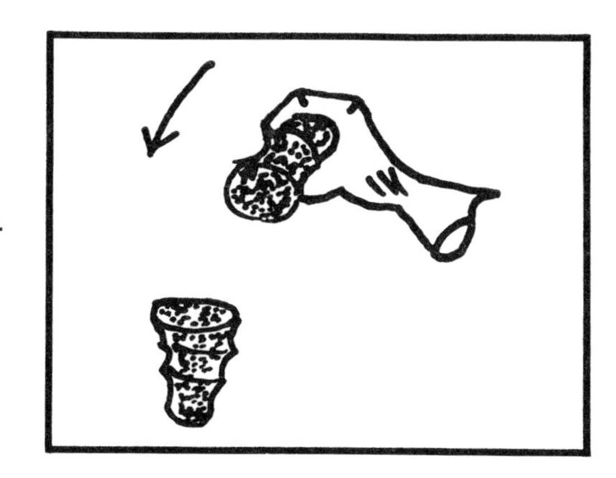

Fig. 2-5. Shake test.

Dry Strength

A mold must not only hold it's shape in the green state, it must also hold it's shape in the dry state. This is an important property and is measured as dry compression by allowing the test specimen to dry out before testing which is then carried out in the same manner as for green strength. A good average figure is thirty pounds per square inch. Dry strength should be no higher than necessary, excessive dry strength results in a critical sand. If the molding sand has too high a dry strength it will not give or break down as the casting shrinks during solidification. This will cause hot tearing of the casting. This and other defects are explained in Chapter 11.

Durability

This is the measure of the sands ability to withstand repeated useage without losing it's properties and to recover it's bond strength after repeated useage. The sands fineness, the type and amount of clay bond determines the sands durability. The ability of the bonding clay to retain it's moisture is also an important factor.

Moldability

This characteristic is also related to the nature of the bonding clay and the fineness of the sand.

Because the base sand determines the resulting finish of the casting, it should be selected with care keeping in mind the type, weight, and class of casting desired. The three or four types of screened sands formerly used for a base has given way to the practice of blending one coarse sand with a fine which results in a better grain distribution. This has been found to produce a better finish and texture. Each of the two sands selected should have a good grain distribution within itself. Contrary to popular belief, additives of an organic or carbonaceous nature do not improve the finish but only furnish combustibles resulting in better peel.

Blended Sands

Use the following suggested sand blends as a guide in selecting your own basic blends. Start with a good high grade silica, washed and dried, screened and graded. Adjustments can be made to the base after sufficient tracking is done.

- For heavy iron using green or dry sand. Fineness 61 and 50. Permeability 80 to 120.

- For medium iron using green sand.
  Fineness 70 and 45. Permeability 50 to 70.
- For light squeezer iron, green sand.
  Fineness 110 and 80. Permeability 20 to 30.
- For stove plate iron, green sand.
  Fineness 200 and 160. Permeability 9 to 17.
- For heavy green steel, green sand.
  Fineness 60 and 35. Permeability 140 to 290.
- For heavy steel, dry sand.
  Fineness 55 and 40. Permeability 90 to 250.
- For light squeezer malleable iron.
  Fineness 130 and 95. Permeability 20 to 40.
- For heavy malleable iron.
  Fineness 80 and 70. Permeability 40 to 70.
- For copper and monel.
  Fineness 150 and 120. Permeability 30 to 60.
- For aluminum.
  Fineness 250 and 150. Permeability 6 to 15.
- For general brass.
  Fineness 150 and 120. Permeability 12 to 20.

SAND ADDITIVES

As I have repeatedly cautioned, keep the types, kinds, number and amounts of sand additives to a minimum for best results. However, just so you will be able to judge for yourself when the time comes, I will describe the nature and use of the most popular additives available.

Kaolinite

This is a hydrated aluminum-silicate clay also known as fire clay, it is one of a family of three distinct minerals having a similiar composition, Kaolinite, Nacrite, and Dicite. It is a sedimentary clay of low flux content. It's prime use in molding sands is in facing for very heavy work and also in dry sand work when excellent hot strength is required. It is often blended with western bentonite for this purpose. It comes closer than any other type of bonding clay to producing a sand which approximates the properties of natural bonded sands. It's tough and durable and easy to use.

Bentonite

Both southern and western bentonites are very useful to the foundry and are closely related chemically, the basic difference in their properties is that southern bentonite will impart less baked, hot and dry strength to a mix, it will collapse quicker and has a shorter life requiring larger and more frequent additions to maintain the sand at a fixed level. It seems to impart more green strength part for part. Also is beneficial in the shake out due to its lower hot strength. Western bentonite is more widely used in steel sands and sands for heavy castings requiring high hot strength to prevent cutting and washing. The smart foundryman uses each to its best advantage and blends the two together to give him the properties he desires. It is very common to use 50 percent western and 50 percent southern as the bonding medium.

Sea Coal

This is ground bituminous coal marketed in various grades. Its basic use is to improve the finish, by reducing burning in, creating a reducing atmosphere. The correct amount depends upon the base sand and weight of casting being produced. When the correct amount is used it produces excellent results. As with any sand additive, excessive amounts are detrimental, producing rough castings, fins and veining. It should be remembered that sea coal when heated forms coke increasing its size considerably. When used in amounts of from 4 percent to 6 percent mold hardness is increased somewhat and more moisture is required. The permeability goes down as does the flowability and hot strength. There is a gain in dry strength and green strength. Expansion is decreased and refractoriness is increased. Although not commonly used in brass and bronze sands, one part airfloat sea coal and nine parts of brass sand, make one of the best brass facing sands possible to produce.

Pitch and Gilsonite

Both of these materials are used to produce dry and hot strength along with good hot deformation and as an expansion control. Also used as a binder for dry sand work and black sand cores, it is a baking type of binder. Most prepared black core binders sold today under various trade names are usually a mixture of pitch, dextrin, liglin sulphite, and resin and in order to use these correctly you should know the percentage of pitch. Light work pitch is used from 1/2 to 2/5 percent by weight and when using pitch in dry sand work it must be worked well on the wet side. It is used in skin dry and dry sand work for large aluminum bronze work, dried overnight. Sometimes it is used in the combination of 50:50 with sea coal and also with Southern Bentonite. When used as a replacement to sea coal, it can be used as low as ½ percent to as high as 4 percent by weight. The mold hardness and toughness and green strength increases as does the hot and dry strength. The permeability decreases somewhat and little or no effect on the flowability is encountered. It should be noted that other materials are also used as ingredients in facings such as liglin sulphite, rosin, graphite, coke and fuel oil. Gilsonite is a natural pitch much more potent than coal tar pitch and usually is used sparingly in the order of ½ percent to 1 percent in green sand facing mixes. It is also used in the manufacture of some mold and core washes.

Cereal

This is both a bonding material and a cushion material. The most widely used in the foundry is corn flour produced from wet milling of corn starch. The effectiveness of corn flour depends entirely, as with other sand additives, on how it is used. In synthetic sands it works well from ½ percent to 2½ percent by weight. It is widely used in steel sands. The author believes in using corn flour in non-ferrous sands to impart resilience. Corn flour increases dry compression strength, baked permeability, toughness, mulling time, moisture pick up, mold hardness and deformation and in turn decreases the green permeability, sand expansion, hot compressive strength and flowability. Cereals are replaced with wood flour to some extent, however each has its distinct use. Corn flour is very beneficial when you have weak sands, and no other additive can eliminate or reduce scabs, rattails, buckles and similar defects quite as well as 1 percent to 1½ percent corn flour added to the facing. It also keeps spalding to a minimum.

Wood Flour

All wood flours are not the same, foundry wood flours should be a wood flour with the largest portion of resinous material removed, and possessing a low ash content. The normal percentages used in natural and synthetic sand mixes usually run up to 1½ percent. In core mixes use ¼ to ½ percent cereal and ½ percent wood flour. Wood flour additions will reduce volume changes, hot strength permeability and dry strength, while it increases green compression strength, mold hardness, moisture, deformation and density.

Silica Flour

The most widely used inorganic additive is silica flour. It is used in steel sands to control and increase the hot, dry and green strengths. With the advent of the use of finer base sands for steel, silica flour is used much less now than in the past.

The following weights and measures of the most popular additives will be found useful in making your own mixes.

Pitch ....1.3 lb per quart
Sea coal ....1.5 lb per quart
Corn flour.......................................1.2 lb per quart
Goulac ....2.5 lb per quart
Linseed oil......................................1.9 lb per quart
Gilsonite ....1.3 lb per quart
Bentonite ....1.5 lb per quart
Fire clay ....2.0 lb per quart
Silica flour ....2.2 lb per quart
Silica sand ....3.0 lb per quart
Wood flour......................................10 oz per quart

Natural bonded sands can be purchased from all foundry supply houses and are for the most part rather cheap, usually the freight is in excess of the sand cost. However, natural molding sands can be found for the taking in a great many parts of the country. When I operated a brass and Iron shop in New Orleans, Louisiana, we made all our brass castings in river sand, from the river bank, and the iron castings in sand from the banks of the Red river near Alexandria, La. I worked in a stove works in Anniston, Alabama, when I was hoboing that used sand from a pit not over a half block from the shop. We made some of the sweetest castings you ever saw.

French sand was for many years considered the finest natural bonded molding sand for fine art work, plaques statuary etc. Many of Rodan's pieces were cast in French sand. This sand comes from Fontenay des Roses near Paris. I have been told that they have been digging sand from this pit for some 400 plus years, which I have no reason to doubt. It was common practice for many foundries to keep a ½ ton or more of French sand in the shop for that special job, or a gravemarker casting. Using it as a facing, riddled against the pattern ¼ inch thick then backed with regular or system sand.

Of ten random samples French sand showed the following test average:

Clay	17.2%
Permeability 18 at 7.2% moisture	
Grain Fineness	176
Green Strength	10.5
Shear	3.1

Yankee sand which replaced French sand during World War I comes from around Albany, N.Y. An average of ten samples showed:

Clay	19.8
Permeability	18.4 at 6.5% moisture
Green Compression	16.7
Shear	3.8

A definite set of rules or table of figures as to exactly what would be best in every case would be foolhardy to say the least. I have seen castings up to 300 pounds in brass poured in shop "A" with a permeability of 13 and in shop "B" the same class of work, with a permeability of 9. And, still another shop pouring light weight work of not over 3 pounds in a sand with a permeability of 18.5, a real open sand. Confusing isn't it? The figures given earlier as a base are a good starting point.

Although some call it a science, sand casting in green sand molds is still an art and some problems are never solved.

A case in Point. A large foundry in New Orleans that was started in 1929, a very good shop doing a wide range of work for many years, took on a routine job that they had done many times before with great success, of casting some large simple cast iron cylinder sleeves. Twenty five were cast and looked good. When the customer machined them they looked like swiss cheese inside. The shop cast some 75, all Swiss cheese, then coughed up the job. The competition cast 50 all good on the first go round, the next 50 were swiss cheese. The job went back to the first foundry who cast them for a period of 2 years without a loss. True story, what happened? I was there and came up with all kinds of reasons but who knows?

SAND MIXES

Before hunting for molding sand here are some sand mixes and dope that I have found will serve you well as they have me for some 35 years.

Half & Half

Brass sand for general all around work.
50% Albany 00 or similar.
50% 120 mesh sharp silica.
4.5% Southern bentonite.
4.5% Moisture.
2% Wood flour 200 mesh.

72 Hour Cement Sand

MOLD first day—Wash second day—Pour third day.
10 parts silica sand.
1 part H-Early cement.
1% by weight 200 mesh wood flour.
Oil patterns to obtain a parting when molding.

On the 72 hour cement sand the reference to wash the second day refers to a core wash which you will find in Chapter 6, as this type of mold would be considered a core sand mold. When you get into molding and core work this will start to make some sense.

SYNTHETIC MIX FOR LIGHT BRASS

160 mesh wash float penn silica.
4.5% southern bentonite.
2% mesh wood flour.
4.5% moisutre.

SYNTHETIC MIX FOR HEAVY & MEDIUM BRASS

Nevada 120 silica
4.5% southern bentonite.
2% 200 mesh wood flour.
4.5% moisture.

Any reference throughout this book to synthetic molding sand or mixes simply means that the molding sand was not found already mixed with the required clay bond by nature. Actually the term synthetic is a misnomer as you can see the sand and bentonite are both found in nature, together and separately. A synthetic sand mix merely means the foundryman has purchased a washed and dried silica of the grain size and distribution he feels best suited for job at hand, and added the type and kind of binder needed.

Core mixes are usually made up of sharp silica sand and a suitable binder as you will see—but unlike synthetic molding sands are not called synthetic core mixes.

REBOND FACING

3 parts air float sea coal.
1 part Goulac.
10 parts molding sand.

Now we have a case where our molding sand, floor sand, or system sand (same thing), has become weak from use. The green compression has dropped to a point where it's becoming difficult to mold. We can, if we wish, mix into the sand 3 parts (three shovels, buckets etc.) of sea coal and 1 part (shovels, buckets etc.) of Goulac with each 10 parts (shovels, buckets, etc.) of our weak sand, and in doing so, we revive our floor or system sand back to something we can work with.

You can revive a dead sand heap by sprinkling southern bentonite over the sand molds after they have been shaken out prior to re-tempering and conditioning for the next run. It is done like you would feed chickens by hand.

Rather than try to revive a large amount of sand which is lots of work, a more common practice is to revive only 10 parts of our sand as the formula reads, and face each mold with this and back with the weak sand. In no time the entire system is rebonded. The practice of facing is covered in detail in Chapter 7. Rebonding molding sand over three times is not good pactice. Some shops face all molds with new sand and back or finish the mold with floor sand, the theory being that this continuous introduction of new molding sand keeps the floor sand in good shape. This, however, causes a continuous growth of the floor sand and in no time you are up to your ears in sand and have to get rid of some.

SYNTHETIC MIX FOR BRASS & BRONZE

AFS fineness 60 to 80—98.5% silica content

95.5% by weight dry sharp silica.
0.5% corn flour.
4.0% 50/50 Southern Western Bentonite.
1.5% Dextrine.

DRY SAND MIX FOR BRASS AND BRONZE

Use system sand tempered with glutrin water, one pint of glutrin to five gallons of water. Bake till dry at 350°F (5% pitch can be added to the facing sand).

DRY SAND MOLDS FOR PHOSPHOR BRONZE

System sand such as Albany or a synthetic sand with the addition of 1 part wheat flour (bakery sweepings) to each 40 parts molding sand. Temper with clay water 30° Baume (bonding clay—red clay or blue clay, not fire clay). For aluminum bronze leave out flour if desired.

CEMENT SAND MIX FOR LARGE BRASS & BRONZE

6.5% moisture.
12% Portland cement.
81.5% sharp silica sand approximately AFS 40 fineness.

FACING FOR PLAQUES AND ART WORK

5—shovels fine natural bonded sand (Albany or equal).
¼—shovel powdered sulphur.
¼—shovel iron oxide.
Spray mold face with molasses water no later than 10 seconds after drawing pattern (this is one of the most important steps in producing a plaque).

The molasses water in this case consists of 1 part blackstrap molasses to 15 parts of water by volume. Allow the mold to air dry a bit before pouring.

Brass sand can be made up synthetically or part synthetic—part natural sand by choosing the correct base sand for the class of work and bonding with 50:50 south and west bentonite. A very popular all around mix is:

BRASS MIX

90 lb AFS 90–140 grain silica.
10 lb natural bonded sand 150–200 fine.
4 lb southern bentonite.
1 lb wood flour.

To make any non-ferrous mold suitable for skin dry work or dry sand work, simply temper facing with glutrin water or add up to 1% pitch to the facing.

HI NICKEL FACING SAND

15 parts silica AFS G.F. 120 and ½ part bentonite.
3 parts system sand and ½ part fire clay.

For dry sand molds a good all around sand for brass and bronze is;

BRASS AND BRONZE MIX

dry new sand—95.5 pounds
bentonite—3 pounds
corn flour—0.2 pounds
dextrine—1.3 pounds

In general for aluminum you can use a much finer base sand, synthetic or natural bonded sand, due to the low melting and pouring range of aluminum. As a guide line, the two sands given here would cover 90 percent of all aluminum casting work.

Aluminum Mixes

Natural bonded sand

grain fineness	210 to 260
clay content	15 to 22%
moisture	6%

This should give a green strength of 6 to 8 and a permeability of 5 to 15

Synthetic

grain fineness (washed and dried silica)	70 to 160
southern bentonite	4 to 10%
moisture	4 to 5%

When you compare these two basic molding sands you will note that the synthetic sand is more open with a resulting higher permeability with a lower moisture content. This sand, in areas of low humidity, is hard to keep wet enough to mold with due to evaporation of the already low moisture required. If you lose say 2 percent of the moisture on a sand carrying 4 percent you dropped 50 percent. Lose 2 percent of your 6 percent moisture with a natural bonded sand and you have dropped only 33 1/3 percent which will not materially affect its moldability.

The synthetic sand can be rammed harder without casting blows due to its low moisture and high permeability but will produce a casting surface rougher than its sister sand (natural bonded).

So what have we here, natural versus synthetic. Not really, each has its place. In order to make up synthetic sand you must have a sand muller in order to mix the bond and sand together. It must be mulled each time it has been used. A muller is too expensive for a small one or two man shop. Natural bonded molding sand can be conditioned with a molder's shovel. I would be amiss not to cover synthetic sands to the extent that the reader is afforded some knowledge about them. The fact remains that man has been molding for many thousands of years with natural bonded molding sand.

Natural bonded sands sold by foundry supply houses can be purchased in grades suitable for casting very small light castings to heavy castings. Most are compared to Albany grades and classified similarly. Albany natural bonded graded sands which have been the standard for years are classified as follows.

Albany Molding Sand
No. 00 AFA classification No. 1E (AFA-American Foundryman's Association)
Use: Very small iron, brass and aluminum
No. 0 AFA classification No. 2-D or E
Use: Medium weight castings where a smooth surface is essential.
No. 1 AFA classification No. 3-D
Use: Medium weight iron & brass General bench work.
No. 1½ AFA classification No. 3-D
Use: Medium to chunky iron & brass
No. 2 AFA classification No. 4-D
Use: Medium iron-heavy brass
No. 2½ AFA classification No. 4-D
Use: Side floor molding and medium heavy iron & brass
No. 3 AFA classification No. 5-D
Use: Heavy Iron and dry sand work.

CONDITIONING THE SAND

The most rudimentary method used to put natural bonded molding sand into the proper condition suitable for molding is with a molder's shovel.

The molder's shovel is 38 inches long with a 9 inch by 12 inch flat blade. The handle is wedge shaped into a peen, the actual peen, a rubber piece dovetailed into a cast handle. See Fig. 2-6.

The actual cutting of the sand consists of first windrowing the sand into a long low row very much like a ridge in a plowed field ready for planting. See Fig. 2-7.

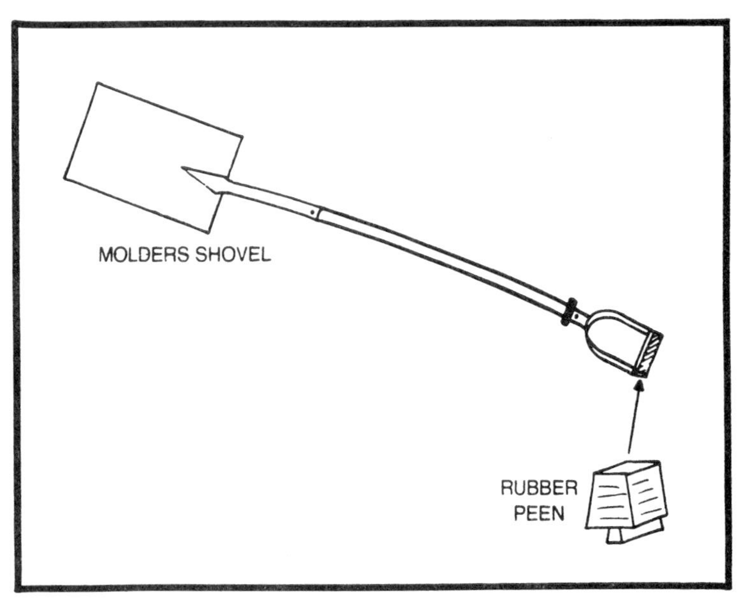

Fig. 2-6. Molders shovel.

The windrow is wetted (tempering) down with a regular gardener's or molder's sprinkler can. The object is to work the moisture evenly into the sand and break up any lumps (cutting). Starting on one end, the windrow is hit a sharp blow with the back on the shovel blade. See Fig. 2-8.

Scoop up the portion of sand which has been struck and turn it over and strike it again. Do this the entire length of the windrow. Now sprinkle the windrow lightly all over. Now, starting at one end turn over a shovel full (each time you turn over the sand the shovel is scooped completely under the windrow at floor level). Hit it with the back of the shovel a good stout blow, pick up the section you have just struck and with a fanning motion throw the sand into a pile at your molding bench.

Aerating the sand or fluffing it up is accomplished by giving the blade a quick twist while at the same time throwing the sand in an arc. When all the sand has left the blade the blade should be almost vertical to the floor. See Fig. 2-9.

The shovel is held with the left hand as close to the blade as possible with the right hand grasping the handle. The entire motion must be a simple rhythmic action. When you accomplish it you will find that it is not tiring and is good exercise. As to the correct moisture in the sand by your sprinkling, cutting and aerating, this can

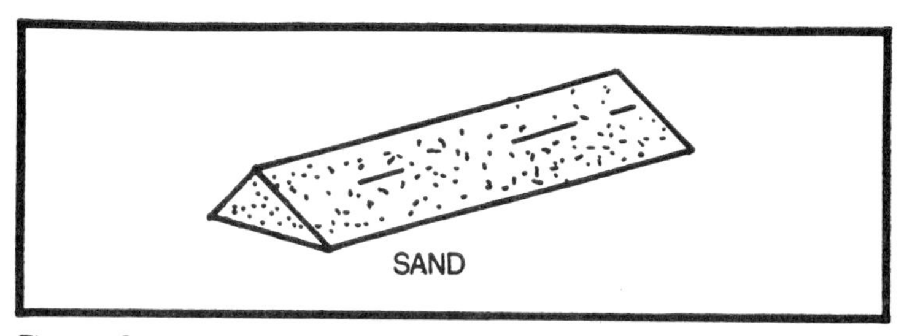

Fig. 2-7. Sand windrow.

only be learned by doing and knowing the feel of a hand full of sand squeezed and broken. (Old timer test.)

You can get some sort of an idea how much moisture to use, by taking 10 pounds of dry natural sand, placing it in a plastic shallow container and slowly add moisture from a container with a known weight of water in it, and mixing it with your hands by rubbing it between your hands between each moisture addition. When it's right weigh the remaining water, then you have something to go on.

If you have 100 pounds of sand dry, 6 pounds of water should give you 6 percent moisture. Start with 5 pounds of moisture. Now, when you cast metal into your sand molds and shake them out, the heat from the metal removes only a percentage of the moisture which must be added when the sand is reconditioned. This percentage varies from quite a lot for a large chunky casting (brass) poured at 2100 degrees Fahrenheit to very little for a small aluminum casting poured at 1450 degrees Fahrenheit. You will soon find an optimum moisture value. The whole proposition really rests on the results you are having with your particular sand and weight range of your castings. Even in a foundry where the sand system is mechanized and mechanically mixed under close controls prior to delivery to the molding stations, a good molder can tell from the feel of the sand whether it will do the job or not.

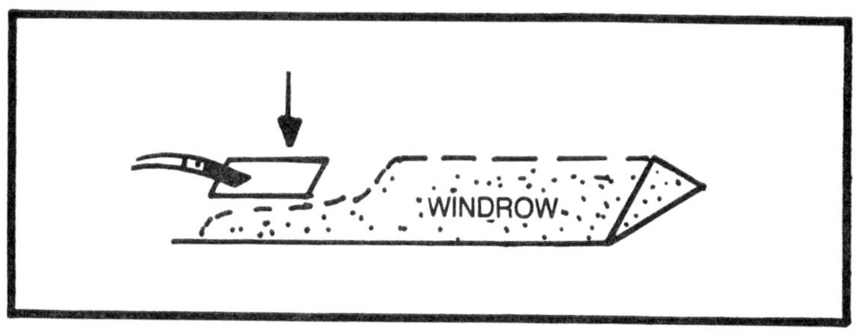

Fig. 2-8. Cutting sand.

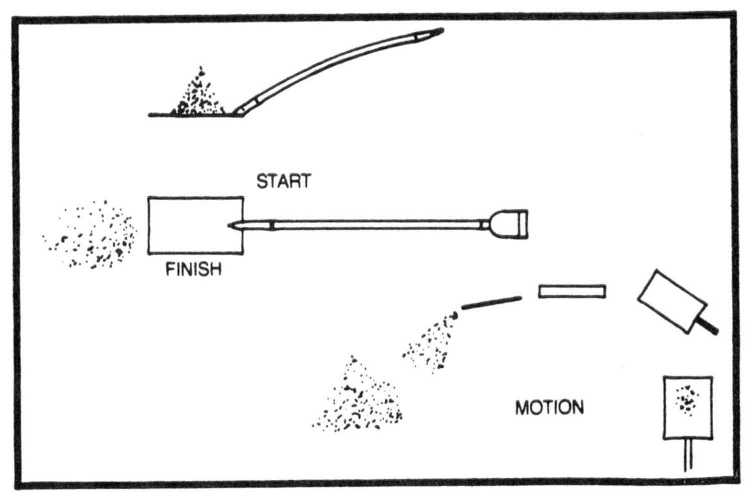

Fig. 2-9. Aerating sand.

The sand must not feel wet or muddy but must have a cool silky feel and good flowability, ram easily into the mold. Slide easily from the shovel without sticking to the blade, nor should it feel crumbly and gritty, or a powdery dry.

Your shovel blade must be kept smooth and clean, free from rust and stuck sand, a good molder will clean his shovel blade at the end of each day and give the blade front and back a heavy coat of floor wax. Molding or conditioning sand with a dirty shovel increases the work load by as much as 200 percent. Visit a foundry and feel the sand. Buy a small amount from a foundry supplier or better yet bum a bucket full from a foundry just to feel and play with.

Keeping The Heap In Condition Over Night

A common practice among small one to five man sand foundries (or a hobby foundry) is after pouring off the molds the sand is simply fanned into the heap and the surface of the heap is patted down smooth with the flat side of the shovel to close up the surface.

The heap is sprinkled all over with water then covered with a sheet of thin builder's plastic and the edges kept down in place with bricks, ingots or what have you. During the night because the moisture and steam from the sand cannot escape or evaporate it will condense on the underside of the plastic and percolate through the entire heap. When uncovered the next morning it's ready to mold. A friend of mine with a one man foundry told me that he has gone away for a 3 week vacation and when he returned his covered heap was still in perfect molding condition.

If you try to temper a pile of dry sand you will appreciate the problem. You can start up a new dry sand heap by windrowing it and giving it a good sprinkling down, then cover it up with plastic and let it percolate until every bit of it is damp, then cut and aerate it into a pile which you keep in shape by keeping it covered tightly.

A big time and work saver for the small foundry is the Combs Gyratory Electric Riddle. This piece of equipment is inexpensive by comparison to all other mechanical sand conditioners available. The Comb Electric Riddles have been manufactured for many, many years. I have one built in 1925 that is still going strong. They are still in business and a brand new one can be bought today. The Combs Riddle is simply an electric riddle or sifter which by its gyratory action tempers and aerates the sand in one operation. The term riddle means to sift through a screen. See Fig. 2-10.

The operation of a Combs riddle is simplicity itself. A one-sixth horsepower low speed motor at its top, drives a shaft which is attached to a fly wheel directly above the riddle. This fly wheel is out of balance by virtue of a lead weight attached to one side of the flywheel's inner rim. When the fly wheel is rotated, it imparts a

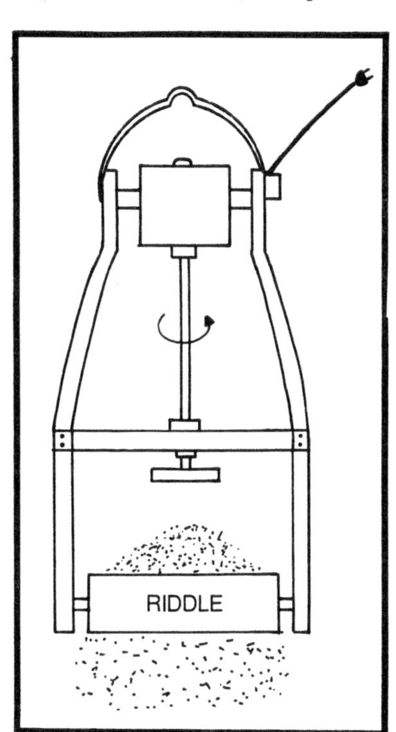

Fig. 2-10. Combs gyratory riddle.

gyratory or shaking motion to the lower screen, causing the sand shoveled into the riddle to mix and go through. The screens are easily removed by a quick clamp arrangement for cleaning or changing. The screens can be purchased in a wide variety of openings from 8 to the inch to ½ inch, the ¼ mesh or 4 squares per inch being most generally used to condition sand,

The riddle will also clean out your sand heap by removing anything that will not go through the mesh, such as tramp metal, rocks and other trash. The 20 inch diameter Combs, the most popular size, can be hung from a tripod type of frame or from the ceiling by a chain. A Combs 20 inch riddle weighs approximately 100 pounds. One riddle suspended between two molders will keep them in sand all day.

Chapter 3
The Molder's Tools

Next to the molder's shovel which we covered in the last chapter, the most used hand tool is the molder's finishing trowel.

The #1 standard finishing trowel comes in three widths 1¼ inch, 1½ inch and 1¾ inch. The blade length is standard for any width, 6 inches long. The 1½ inch width is standard. The blade tapers from 1½ inches wide at the handle end to 1 inch wide at the end of the blade which is rounded. The handle tang comes up straight from the blade 2 inches, then is bent parallel to the blade to receive the round wooden handle. The 2 inch rise gives you knuckle room when troweling on a large flat surface. The only difference between a #1 trowel and a #2 is that the #2 finishing trowel has a more pointed nose. See Fig. 3-1.

Finishing Trowel

The finishing trowel is used for general trowel of the molding sand to sleek down a surface, to repair a surface, or cut away the sand around the cope of a snap flask. Other uses are covered in Chapter 7 under molding.

Heart Trowel

The heart trowel is a handy little trowel for general molding and as its name implies has a heart shaped blade. They run in size from 2 inches wide to 3 inches wide in ¼ inch steps. See Fig. 3-2.

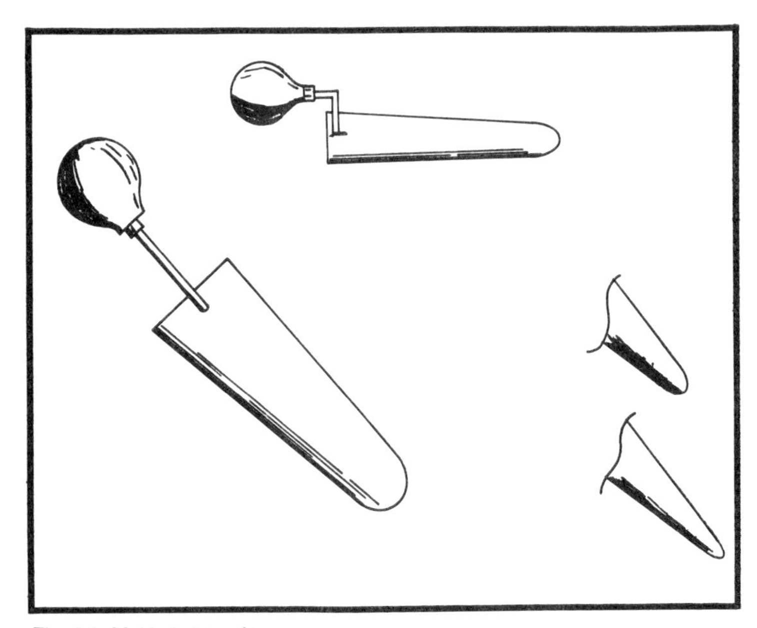

Fig. 3-1. Molder's trowels.

Core Maker's Trowel

The core maker's trowel is exactly like a finishing trowel with the exception that the blade is parallel its entire length, and the nose is perfectly square. They come in widths of from 1 inch to 2 inches in ¼ inch steps and blade lengths of 4½ inches long to 7 inches long in ½ inch steps. This trowel is used to strike off core boxes and due to it being parallel and square ended, a core can be easily trimmed or repaired and sides can be squared up. See Fig. 3-3.

The Bench Lifter

The bench lifter is a simple steel tool with a right angled square foot on one end of a flat bent blade. See Fig. 3-4.

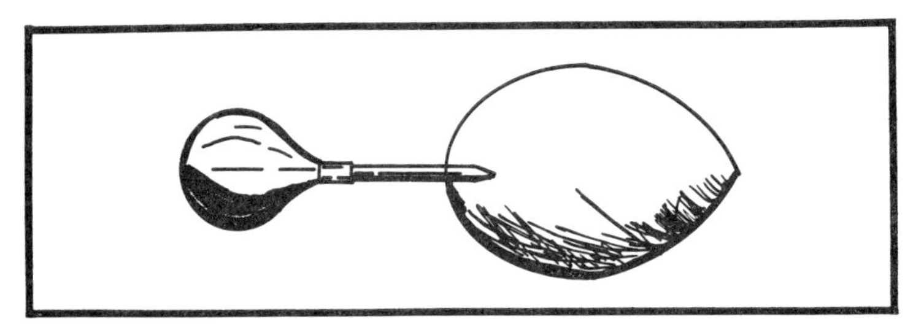

Fig. 3-2. Heart trowel.

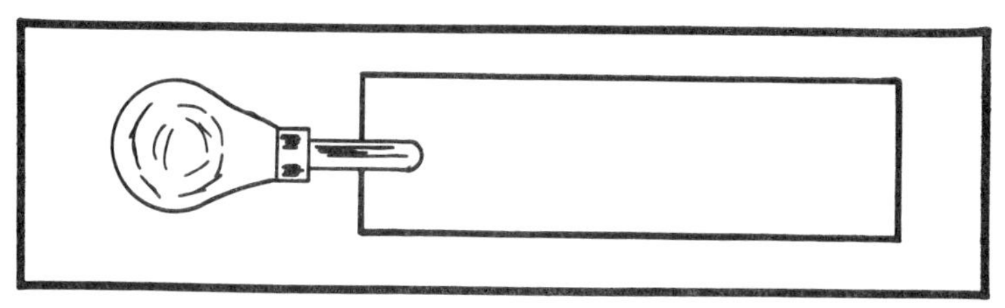

Fig. 3-3. Core maker's trowel.

This tool's biggest use is to repair sand molds and lift out any tramp or loose sand that might have fallen into a pocket. See Fig. 3-5.

To remove dirt from a pocket that will not blow out with the bellows, you simply spit on the heel of the lifter and go down and pick it up, then wipe off the heel and go back and slick the spot down a bit.

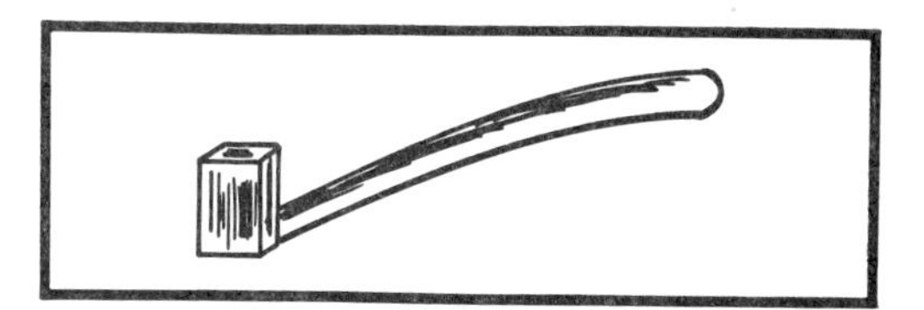

Fig. 3-4. Bench lifter.

Lifters come in all sizes from a bench lifter ¼ inch wide by 6 inches long to floor lifters 1 inch wide and 20 inches long. All have the same use. Many molders make their own lifters to suit the class of work they generally do. Bench, Floor, Pit etc.

Slick and Oval Spoon

This tool is a must for all molders. Again the size needed is determined by the work involved. Most molders have at least 4 sizes

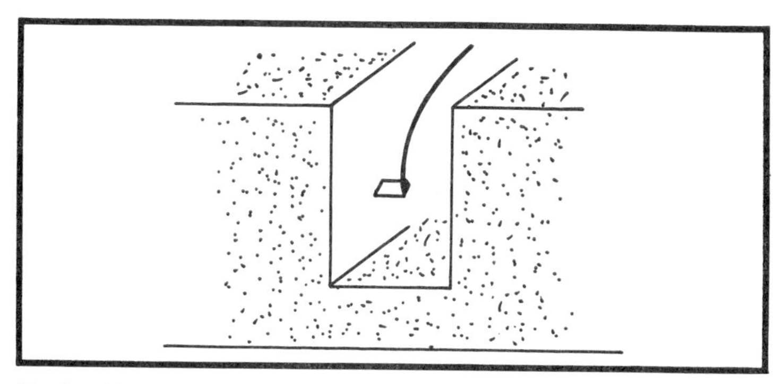

Fig. 3-5. Using lifter.

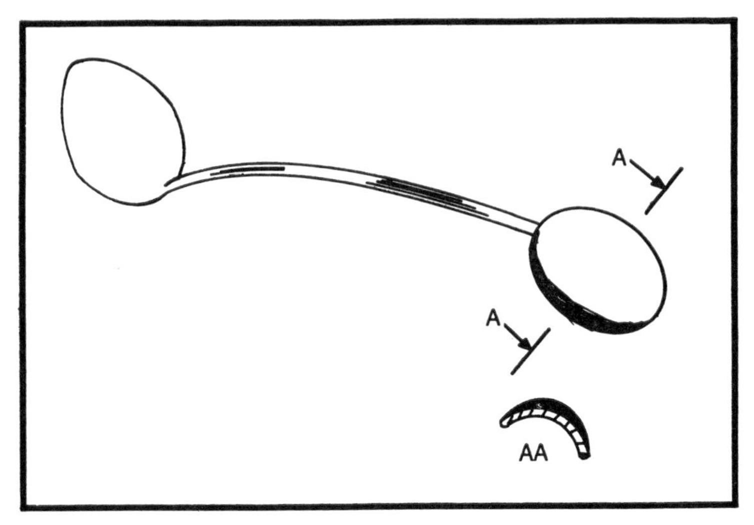

Fig. 3-6. Slick and oval.

from the little bitty one ¼ inch wide to the big guy 2 inches wide. This tool is called in the trade a double ender. One end is a slick similar to a heart trowel blade but more oval shaped. The opposite end is spoon shaped, the outside or working surface being convex like the back of a spoon. Its inner face is concave. This face is never used and therefore is usually not finished smooth, and when new is painted black. Both faces of the slick blade are highly polished. See Fig. 3-6.

The double ender is a general use molding tool used for slicking flat or concave surfaces, open up sprues, etc.

Sucker

The sucker is not a purchased tool but one made by the molder. It consists of 2 pieces of tubing and a tee of copper or iron. See Fig. 3-7.

It's use is to clean out deep pockets in molds where the bellows and lifter fail or if the pocket or slot is just too dirty to spend the time a lifter would take, or where it would be hard to see what you are doing with a lifter. The operation is very simply, you simply blow through the elbow with an air hose. This creates a vacuum in the long length which gives you in effect a vacuum cleaner with a long skinny snout. Now stick the long end down to where the problem is and blow through the elbow. These jobs will lift out small steel shot, a match stick or material which cannot be wetted such as parting powder or silica sand. See Fig. 3-8.

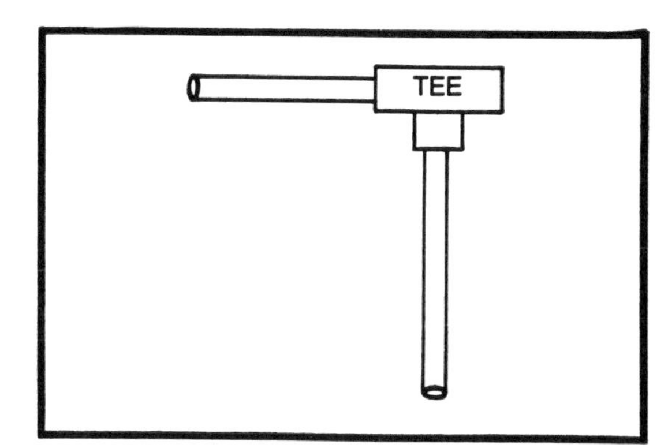

Fig. 3-7. Sucker.

Now a word of warning, watch in what direction you have the discharge end of your sucker pointed, when sucking out a mold. You can blow sand in some bird's eyes who is quite a distance from you and wind up framed with a molding shovel.

Molder's Brush

The most popular size for general bench work and light floor work is the block brush. It is 1 and ⅛ inches wide and 9 inches long. It has four rows of bristles.

It's main use is to brush off the pattern or match plate and the bench, and sometimes the molder himself.

Dust Bags

Two are needed, one 5 inches × 9 inches for parting powder, one 7 inches × 11 inches for graphite. These bags can be purchased from supply houses, but most give them away. You can, as I do, use an old sock (with no holes) for both parting and blacking. The parting

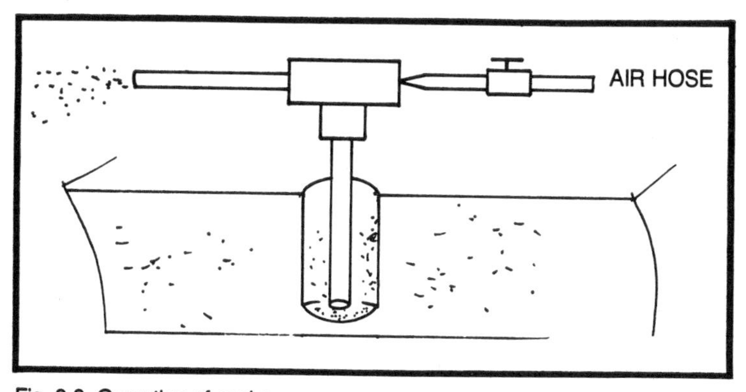

Fig. 3-8. Operation of sucker.

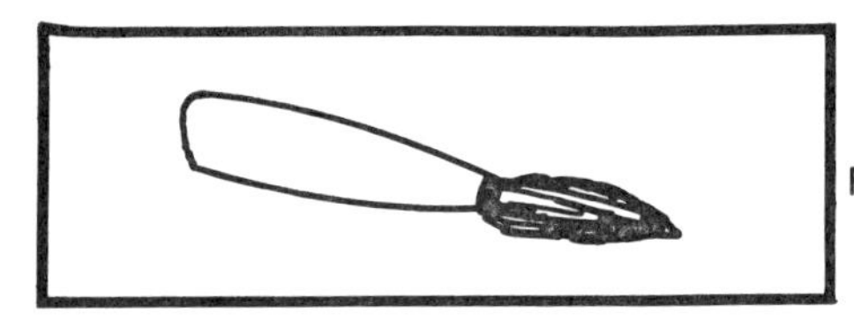

Fig. 3-9. Camel's hair swab.

bag is used to shake parting on the pattern and parting line of the mold faces. The blacking bag is used to dust blacking on a mold surface prior to sleeking.

Camel Hair Swab

A round camel hair brush (swab) used to apply wet mold wash and core wash or blacking to a mold or core prior to drying. Most molders carry two sizes, one ⅝ inches in diameter and one ⅞ inches in diameter. See Fig. 3-9.

Bell Top Wood Sprues and Wood Sprue Sticks

These you can turn or purchase in various diameters used to form sprue and pouring basins in sand molds. In time you wind up with a selection. See Fig. 3-10.

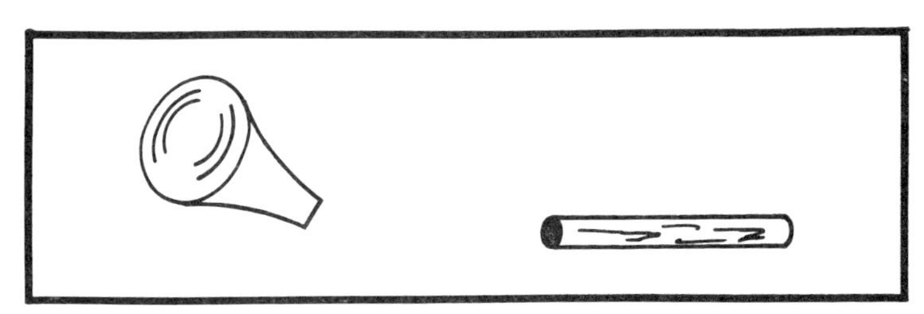

Fig. 3-10. Wood sprues.

Hand Riddle

You need two 18 inch diameter riddles one with #4 mesh and one with #8 mesh, galvanized iron screen. The #4 to be used for general bench and floor molding to riddle the first sand over the

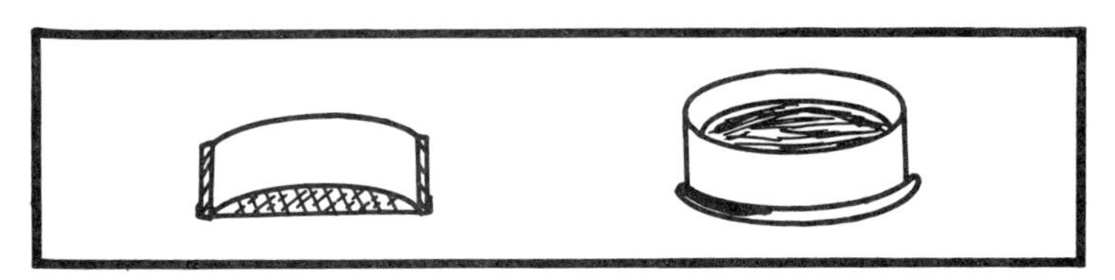

Fig. 3-11. Hand riddle.

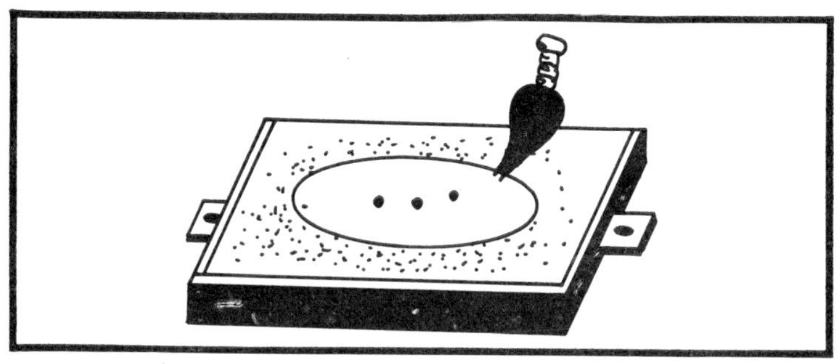

Fig. 3-12. Use of flax swab.

pattern before filling the mold with the shovel and ramming. The #8 is for the same purpose but to apply fine facing sand to special jobs such as plaques or grave markers. See Fig. 3-11.

Flax Swab

Also called a horse tail, it is used to swab the sand around the pattern at the junction of pattern and sand. To dampen this sand to prevent it from breaking away when the pattern is rapped and lifted from the sand, it is dampened by dipping it into a pail of water and shaking it out well. Used on floor work where the pattern presents a fair sized perimeter. See Fig. 3-12.

The swab is also used by some molders to apply wet mold wash or blacking to a mold surface (this requires great dexterity).

Molder's Bellows

In general there are two types, one a short snout or bench bellows and one a long snout floor bellows. The theory is that a

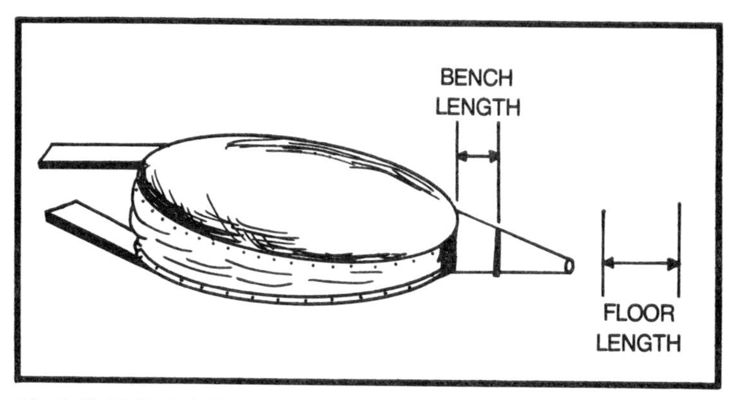

Fig. 3-13. Molder's bellows.

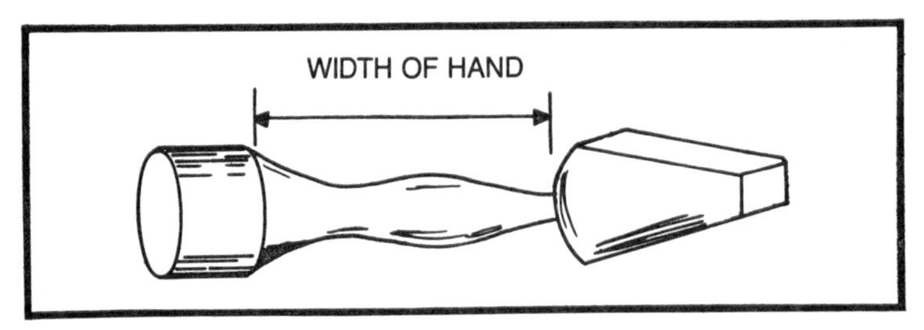

Fig. 3-14. Bench rammer.

molder can wreck a bench mold blowing it out with a long snout floor bellows by hitting the sand with the snout. This is true due to the different stance and angle when blowing a mold bench high, or on the ground. With care one only needs a 9 inch or 10 inch floor bellows to blow out cope and drag molds, sprue hole and gates. See Fig. 3-13.

Bench Rammer

Made of oiled maple. You can buy one or turn one on a lathe, and band saw the peen end with its wedge shape.

The butt end is used to actually ram the sand into the flask around the pattern. The peen end is used to peen or ram the sand tightly around the inside perimeter of the flask to prevent the cope or drag mold from falling out when either half is moved, rolled over or lifted. See Fig. 3-14.

Floor Rammer

Same purpose and design as a bench rammer only the butt and peen ends are made of cast iron and attached to each end of a piece of pipe or a hickory handle. The average length is 42 inches. See Fig. 3-15.

Raw Hide Mallet

Minimum weight 21 ounces used to rap pattern or draw pike to loosen pattern from sand so it can be easily withdrawn from the mold smoothly without damage to mold or pattern.

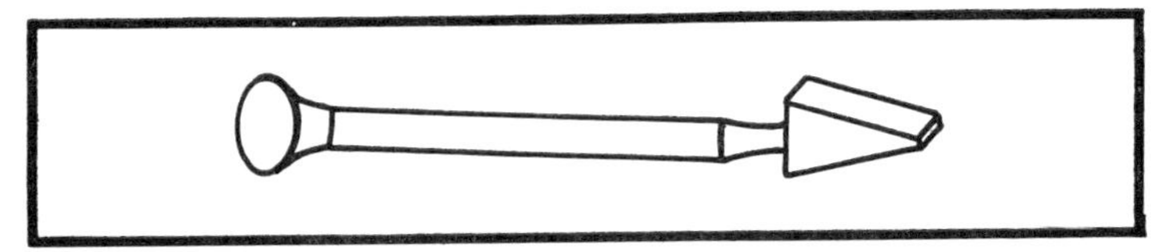
Fig. 3-15. Floor rammer.

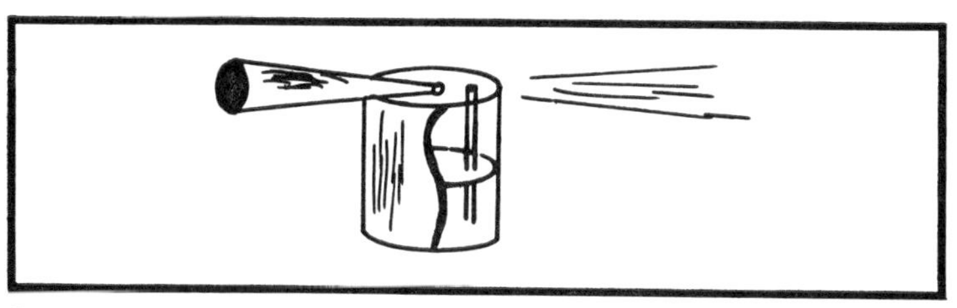

Fig. 3-16. Molder's blow can.

Molder's Blow Can

The blow can is a simple mouth spray can used to apply liquid mold coats and washes to molds and cores, or with water to dampen a large area. It can also be operated with the air hose like the sucker. See Fig. 3-16.

Tubular Sprue Cutters

A tapered steel or brass tube used to cut a sprue hole in the cope half of a sand mold, sold in sizes from ⅞ to 1¼ inches in diameter, all are 6 inches long. See Fig. 3-17.

Draw Pins, Screws & Hooks

These items are used to remove or draw the pattern from the mold. The draw pin is driven into the wooden pattern and used to lift it out as a handle. On a short small pattern one pin in the center will do it. If the pattern is long, use one on each end for a two hand straight lift. The draw hook is used the same way. The draw screw is screwed into the pattern for a better purchase on heavier patterns, and prevents the pattern from accidently coming loose prematurely and falling, damaging the mold, pattern or both. In large patterns, plates are let into the pattern at its parting face which have a tapped hole into which a draw pin with a matching thread is screwed for lifting by hand or with a sling from a crane, two or more are used. See Fig. 3-18.

Metal patterns if they are loose patterns are drilled and tapped to receive a draw pin.

Fig. 3-17. Tubular sprue cutter.

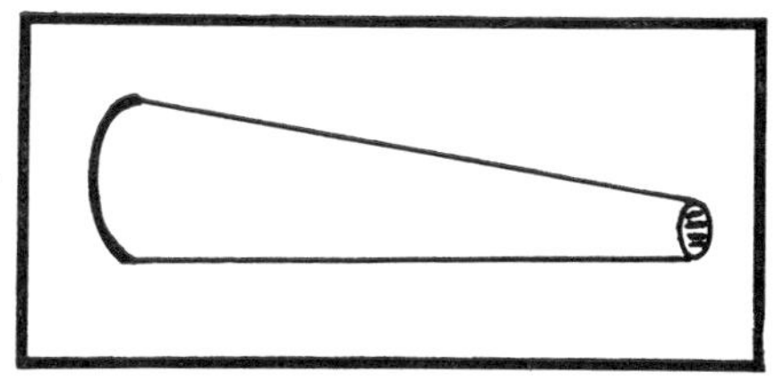

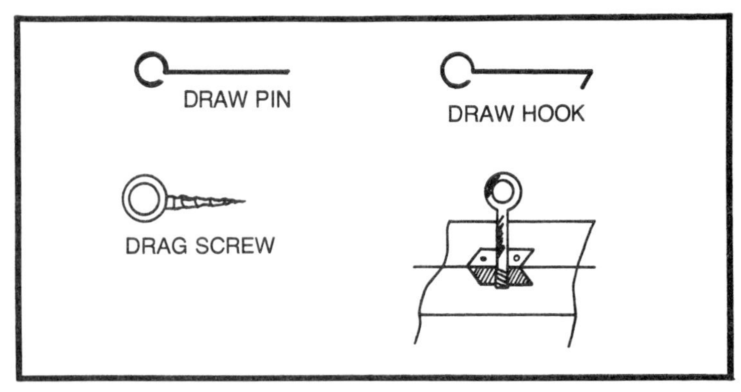

Fig. 3-18. Draw pins and draw plate.

Gate Cutter

The best gate cutter is made from a section of a Prince Albert Tobacco Can. See Fig. 3-19.

To cut a gate or runner in green sand the tab is held between the thumb and first finger and the gate or runner is cut just as you would cut a groove or channel in wood with a gouge. The width is controlled by bending the cutters sides in out, the depth controlled by the operator.

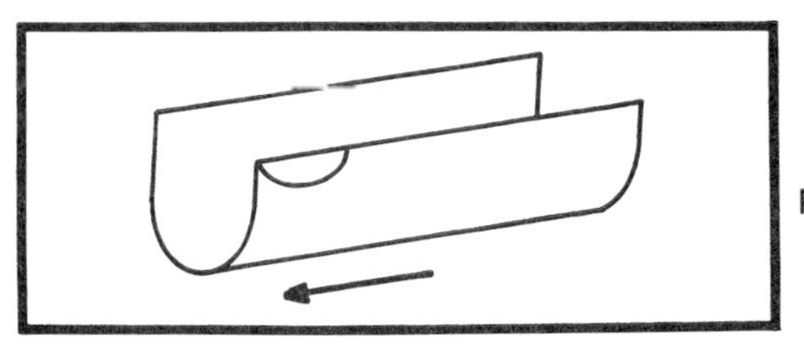

Fig. 3-19. Gate cutter.

Rapping Bar and Rapper

The rapping bar consists of a piece of brass or steel (cold roll) rod which is machined or ground to a tapered point. The rapper is made of steel or brass and is shaped exactly like the frame of a sling shot. See Fig. 3-20.

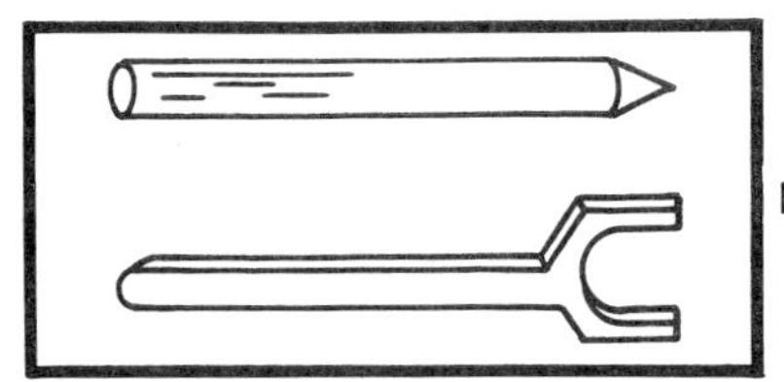

Fig. 3-20. Rapping bar and rapper.

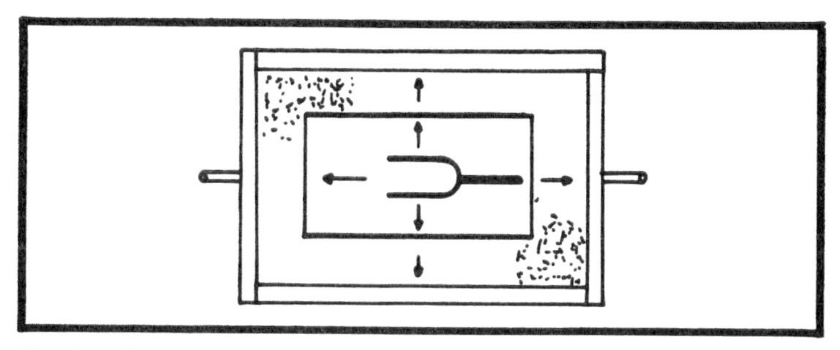

Fig. 3-21. Rapping.

The purpose of these tools is to rap or shake the pattern loose from the sand mold in order to draw it easily from the sand. What you are doing is shaking the pattern in all directions which will drive the sand slightly away from the pattern. The resulting mold cavity will actually be a fuzz larger than the pattern. The pattern should only be rapped enough to free it from the sand. You will be able to see when it is loose all around by the movement of the pattern. Over rapping will distort the mold cavity which may or may not matter depending on how close you wish the casting to hold a tolerence. See Fig. 3-21.

The operation is quite simple, the bar point (of the rapping bar) is pressed down into a dimple in the parting face of the pattern with the left hand. The rapper is used to strike the bar with the inner faces of the yoke.

Bulb Sponge

The bulb sponge consists of a rubber bulb with a hollow brass stem terminating in a soft brush. The stem is pulled out and the bulb filled about three fourths full of water and the stem replaced. The bulb sponge is used to swab around the pattern prior to drawing it, same as the flax swab is used for floor molding. The bulb is gently squeezed to keep the brush wet while swabbing. See Fig. 3-22.

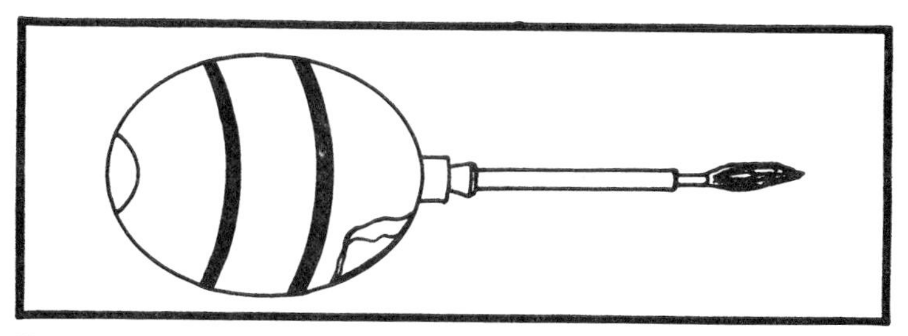

Fig. 3-22. Bulb sponge.

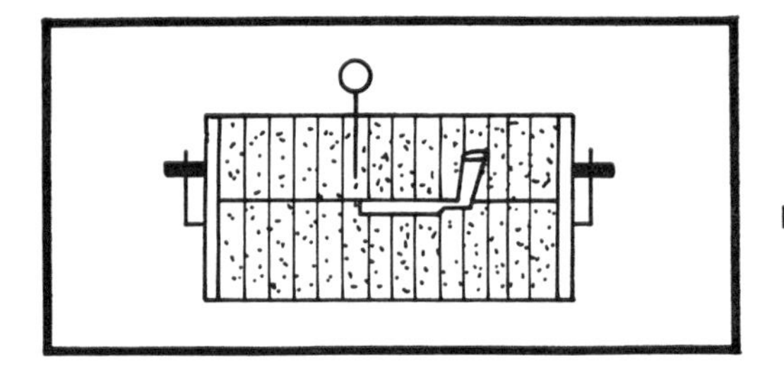

Fig. 3-23. Vent wire.

The Vent Wire

The vent wire is simply a slender pointed wire with a loop at its top, used to punch vent holes in the cope and drag of a sand mold to provide easy access of steam and gases to the outside, during the pouring of the casting. The venting is done prior to the pattern removal. The vent wire is pushed down into the sand to within close proximity of the pattern. The first and second finger straddles the wire each time it is withdrawn as a guide. See Fig. 3-23.

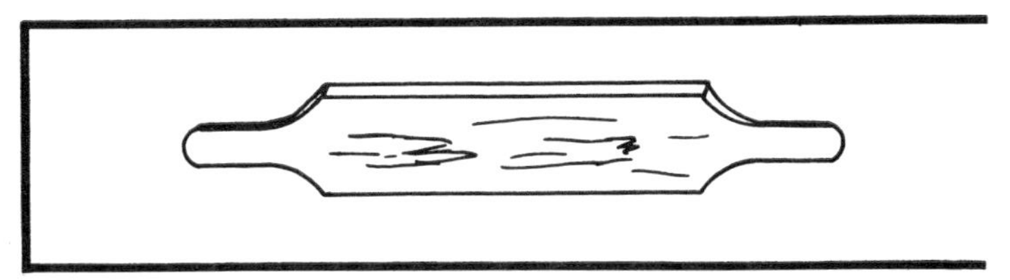

Fig. 3-24. Strike off tool.

Strike Off Bar

Each time a mold is rammed up the sand must be struck off level with the flask both cope and drag. The bar simply consists of a metal or hardwood straight edge of sufficient length. See Fig. 3-24.

Chapter 4
Mold Making Equipment

Sand molds used in metal casting are made in wooden frames called flasks. They have no resemblance at all to the glass or metal liquid container most people think of when they see or hear the word flask.

They actually are open wooden frames that can be held together with pins and guides. They are separated during the mold preparation process and placed back together for pouring the mold without losing the original register. A flask with good pins and guides will close back together in the exact same place every time. This is essential to avoid shifts in the mold cavity with a resulting defect.

BASIC FLASK

The top frame of the flask is called the cope and the bottom frame is called the drag. See Fig. 4-1.

In some molding operations you need one or more sections between the cope and the drag. These sections are called cheeks. See Fig. 4-2.

The pins and guides that are used to hold the sections together can be purchased in a wide variety of types and configurations, round and half round, double round and vee shaped. Single, double or triple vee shapes together with matching guides. Both pins and guides have attached mounting plates by which they can be bolted, screwed or welded to the halves which make up a complete flask set. See Fig. 4-3.

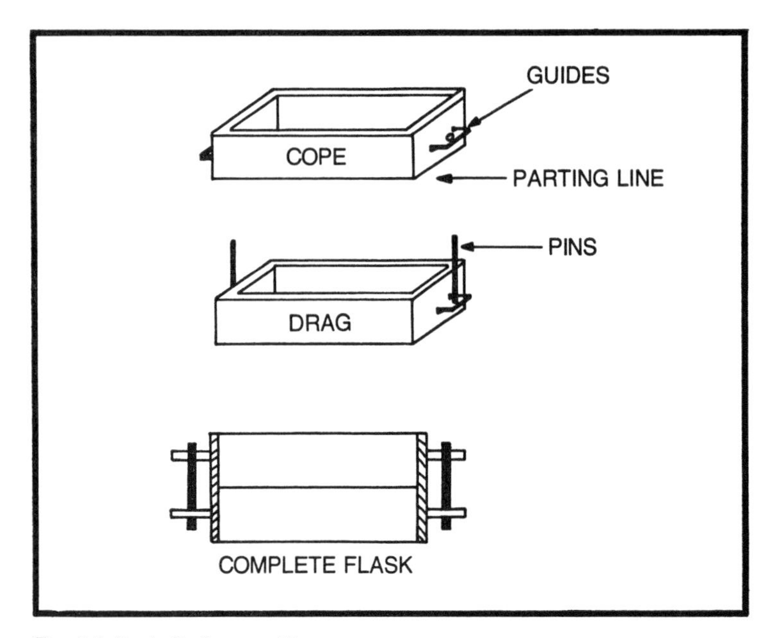

Fig. 4-1. Basic flask assembly.

HOME MADE WOOD FLASK

In hobby or small shops you will find wood flasks with simple wood guides and pins. Although they work for a while they soon loosen or burn up from spilled metal. Only resort to this for a one quick job for which you must build a quick flask to fit it. See Fig. 4-4.

FLOOR FLASKS

These are flasks that are too large to be handled on the bench. These can be of wood or metal. The wood flasks are constructed in such a manner that the long sides provide the lifitng handles for two man lifting and handling.

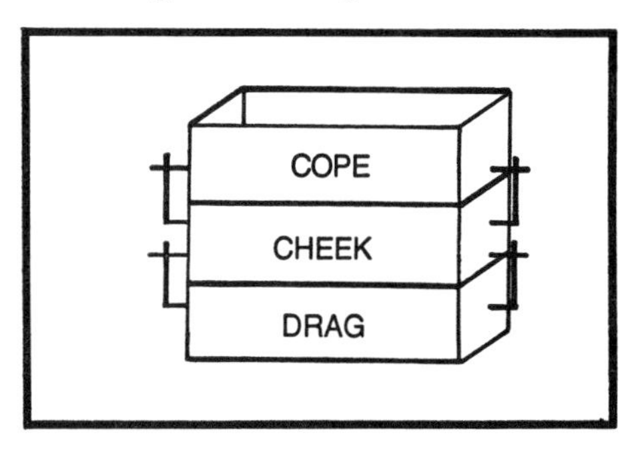

Fig. 4-2. Flask with cheek.

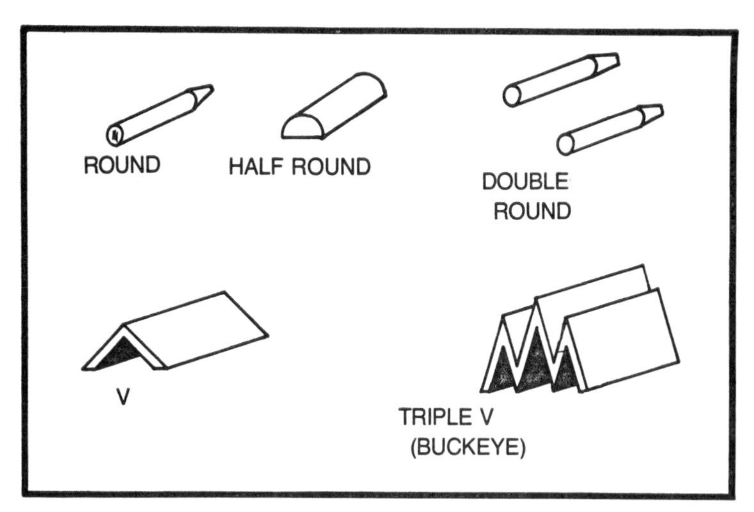

Fig. 4-3. Alignment pin types.

In floor flasks unlike bench flasks when you reach a size of 18 inches × 18 inches to 30 inches × 30 inches members are put in the cope which are called bars. These bars help support the weight of the sand in the cope to prevent it from dropping out. Bars are required in the drag half of the flask only when it is necessary to roll the job over and lift the drag instead of the cope in which case they are called grids. See Fig. 4-5.

The bars do not come all the way to the parting but clear the parting and portion of the pattern that is in the cope by a minimum of ½ inch. These bars in many cases have to be contoured to conform with the portion of pattern that is in the cope. See Fig. 4-6.

In all cases the bars are brought to a dull point along their lower edge to make it possible to tuck and ram the sand firmly under the bars. See Fig. 4-7.

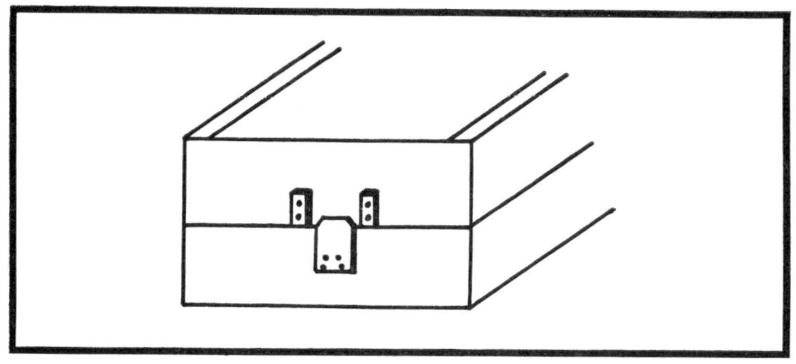

Fig. 4-4. Wood pins and guides.

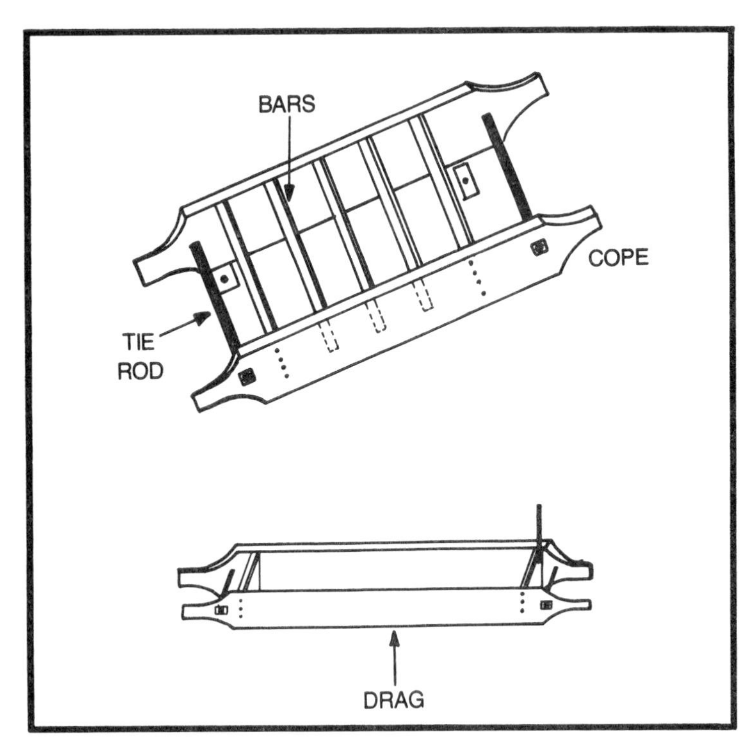

Fig. 4-5. Wood floor flask.

The inside surfaces of both the flask and the bars in the cope section are often covered with large headed roofing nails with the head projecting ⅛ to ¼ of an inch. This gives the entire inner surface an excellent tooth and a good purchase on the sand.

Of course the floor flask like all others must be provided with suitable pins and guides on both ends.

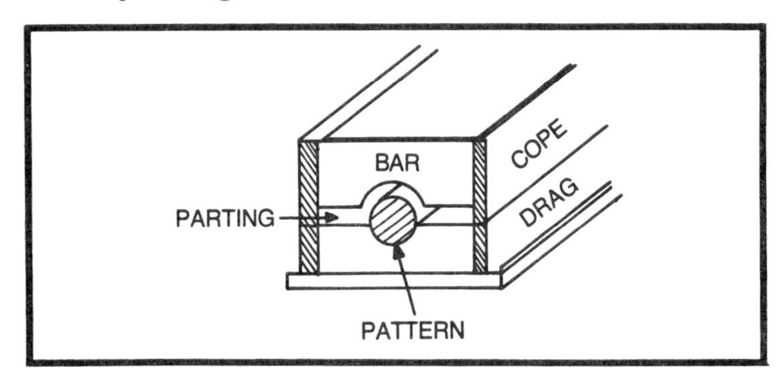

Fig. 4-6. Contoured cope bar.

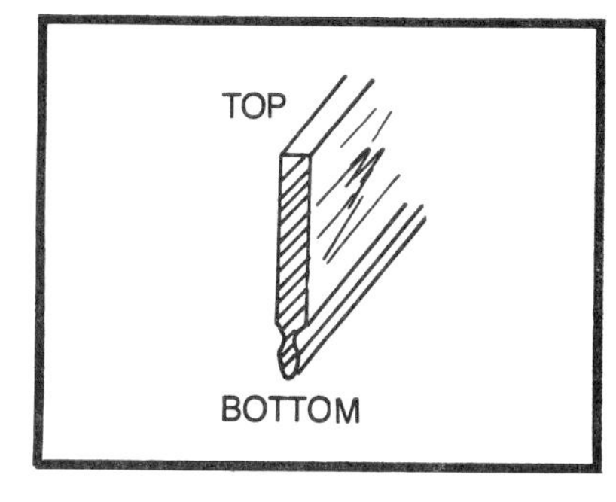

Fig. 4-7. Contoured bar.

Small floor flasks up to 30 inches × 30 inches with a cope and drag depth of from 5 to 8 inches can be made of 1 inch lumber. From 32 inches × 32 inches to 48 inches × 48 inches, 5 inches to 10 inches cope and drag, use 2 inch lumber. From 50 inches × 50 inches to 62 inches × 63 inches 6 inches to 19 inches cope and drag, use 2½ inch lumber. From there up to a flask that measures a maximum of 85 inches × 85 inches with a cope and drag depth of 7 to 30 inches use 3 inch lumber. The bars should be made of the same thickness of lumber as the flask. The number of bars required is generally such that you have a maximum of 6 inches of sand between them.

Now if your flask is extremely wide you might need a cross bar between the bars. This little bar is called a chuck. See Fig. 4-8.

When you get to a flask larger than 85 × 85 inches you should move to welded steel.

Where the type of work permits, the floor flask can be fitted with roll off hinges. These hinges allow the flask to be opened like a book; the pattern removed and the mold closed all without lifting the cope. See Fig. 4-9.

LARGE STEEL FLOOR FLASKS

Large steel floor flasks are more often equipped with female guides on both cope and drag halves with loose pins used for molding

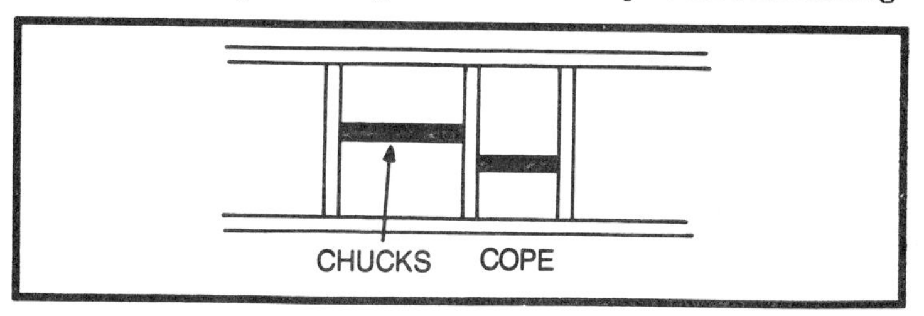

Fig. 4-8. Cope chucks.

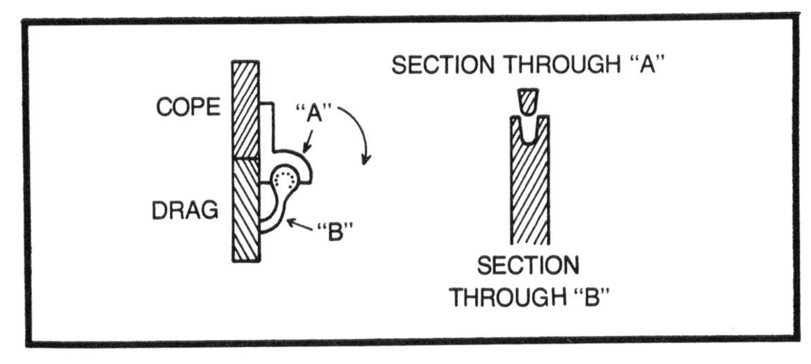

Fig. 4-9. Roll off hinge and guide.

and closing. Some have single hole guides, some double hole guides. See Fig. 4-10.

Steel flasks can be purchased in all sizes with any size cope and drag depth, or combination of different cope and drag depths and any type of pin and guide arrangement your heart desires.

SNAP FLASK

A snap flask is a flask, usually made of cherrywood, whereby after the mold is made the flask can be removed by opening the flask and lifting it off of the mold, leaving the mold as a block of sand on the bottom board. See Fig. 4-11.

Both cope and drag have in corner "A," a hinge and in corner "B" a cam locking device. In operation the cope and the drag locks are closed tight and the mold made in the usual manner. When finished, the locks are opened and the flask is opened and removed from the mold. A typical snap flask hinge and lock is shown in Fig. 4-12.

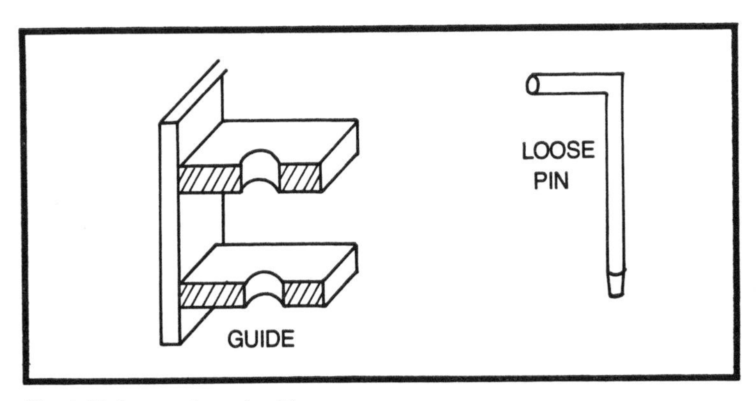

Fig. 4-10. Loose pin and guide.

The big advantage of the snap flask is that you need only one flask to make as many molds a day as you wish. Where with rigid flasks you need as many flasks as the number of molds you wish to put up at a time. I have seen small shops that had only three or four sizes of snaps and a variety of wooden floor flasks.

OTHER TYPES

There are two other types of flasks that are removed from the mold, one is called a tapered slip flask. In this type of flask there is a

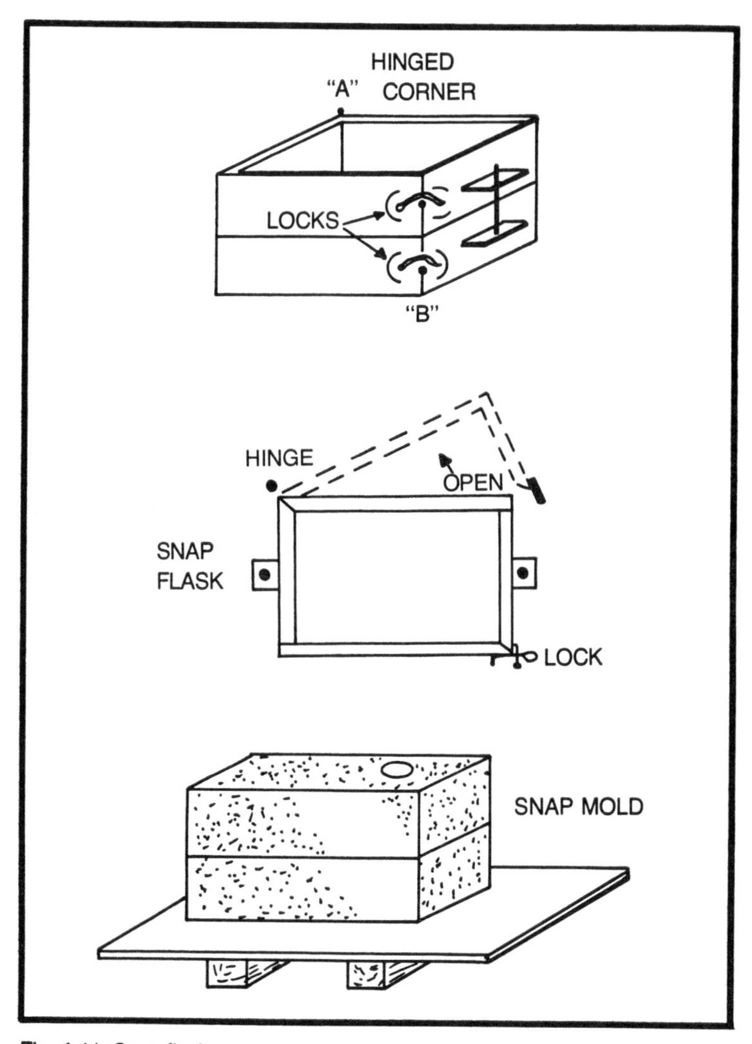

Fig. 4-11. Snap flask.

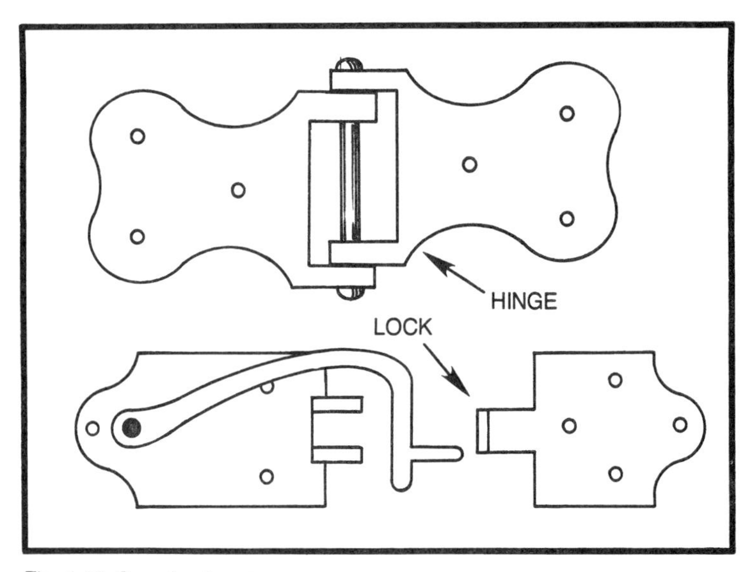

Fig. 4-12. Snap hardware.

strip of metal on the parting face of the cope which prevents the cope from dropping out (while molding) which it would do without the strip, due to the taper on the cope's inner surface. After the mold is made and closed this strip is retracted by a cam lever. When the strip is retracted the tapered flask is easily removed by simply lifting it from the mold. See Fig. 4-13.

The other is called a pop off flask. In this type of flask, both corners cope and drag diagonally across from one another are clamped by a cam locking device. When the mold is completed the cam is released and the corners (which are spring loaded) pop apart

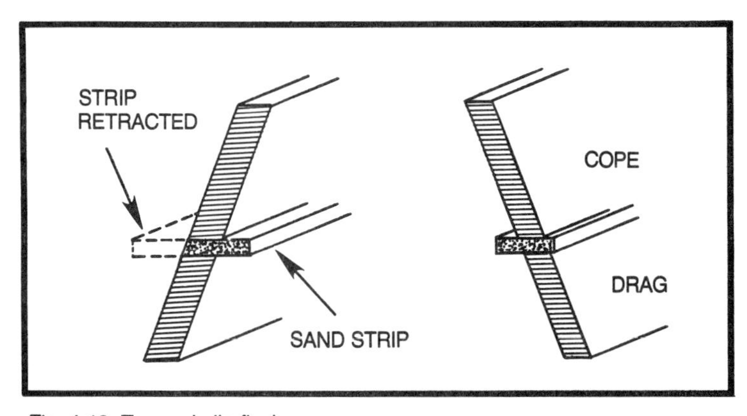

Fig. 4-13. Tapered slip flask.

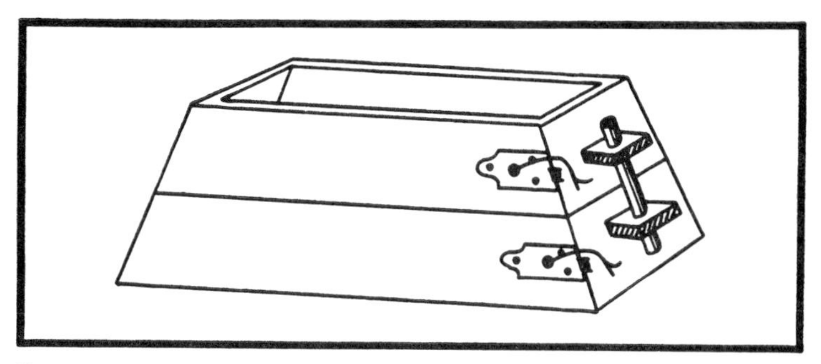

Fig. 4-14. Tapered snap flask.

just enough to release the sand mold. These are quite popular and can be had in enormous sizes for floor and crane work as well as small bench sizes.

JACKET

A jacket is a wood or metal frame which is placed around a mold made in a snap flask during pouring to support the mold and prevent a run out between the cope and drag.

The common practice is to have as many jackets (for each size snap mold) as can be poured at one time. When the molds have solidified sufficiently the jackets are removed and placed on the next set of molds to be poured. This is called jumping jackets. Jackets must be carefully placed on the molds so as not to shift the cope and drag. The jackets must be kept in good shape and fit the molds like a glove. A tapered snap flask and tapered jackets work best due to the taper. See Fig. 4-14.

UPSET

When you need a taller cope section or drag section, then the flask on hand you use a frame of the required additional depth

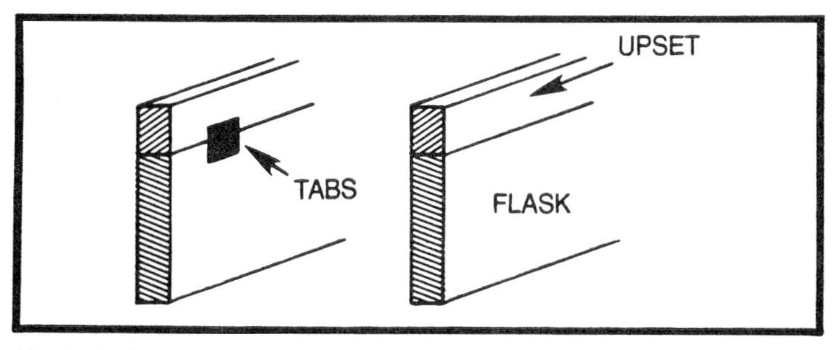

Fig. 4-15. Upset.

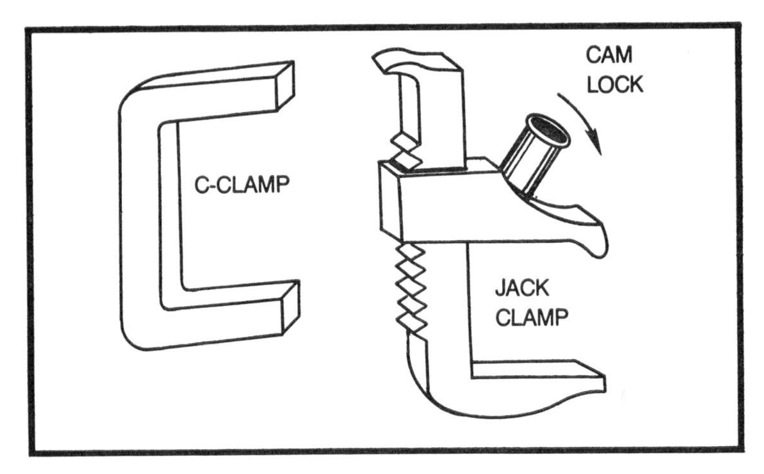

Fig. 4-16. Floor mold clamps.

needed. The inside of the frame is provided with four or more metal strips or tabs which fit down into the flask to hold it in place during molding. These strips or bands (they may be sheet iron bands) are called "upsets" the term "to upset" a cope or drag means to add depth. See Fig. 4-15.

Mold Clamps

Floor molds are always clamped to pour, some times they are weighted, some times not, but always clamped. There are two basic types of clamps used. See Fig. 4-16.

The "C" clamp is cast iron or square steel stock. In operation he bottom foot is placed under the bottom board and a wedge (wood or steel) is tapped between the top foot and the top of the flask side. See Fig. 4-17.

A better practice is to place two wedges between the clamp foot and flask by hand, then tightening them with a small pinch bar, as driving them tight, jolts the mold and could cause internal damage such as a drop. See Fig. 4-18.

The jack clamp is the preferred, but most expensive. Its operation is simple. The foot is placed under the bottom board and the clamping foot is slid down against the cope flask, a bar is placed in the cam lever and pushed down until the clamp is snug and tight.

MOLDING BOARD

The molding board is a smooth board on which to rest the pattern and flask when starting to make a mold. The board should be

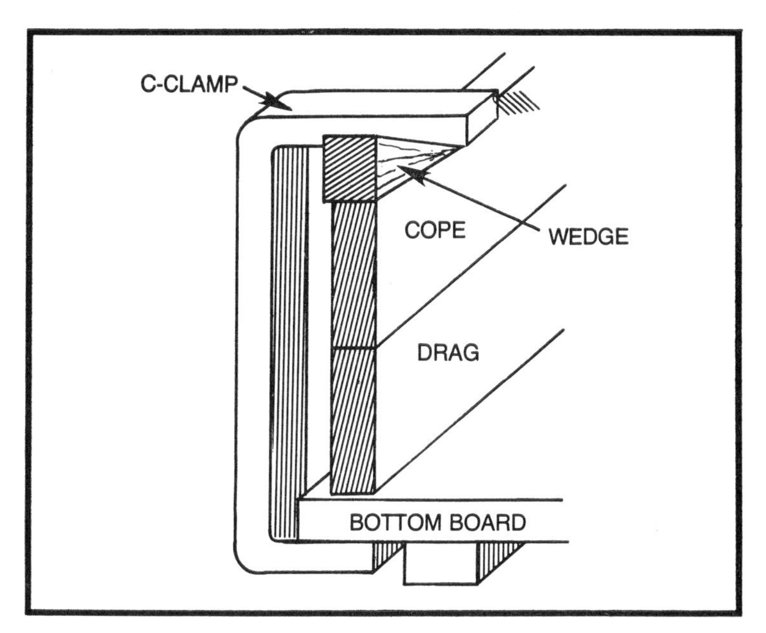

Fig. 4-17. Use of C-clamps.

as large as the outside of the flask and stiff enough to support the sand and pattern without springing when the sand is rammed. One is needed for each size of flask. Suitable cleats are nailed to the underside of the board. Their purpose is to stiffen the board and to raise it from the bench or floor to allow you to get your fingers under the board to roll over the mold. See Fig. 4-19.

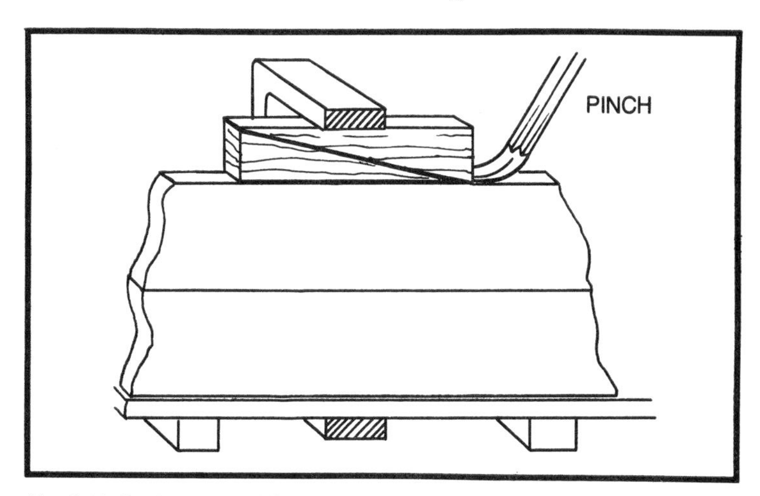

Fig. 4-18. Double wedged C-clamps.

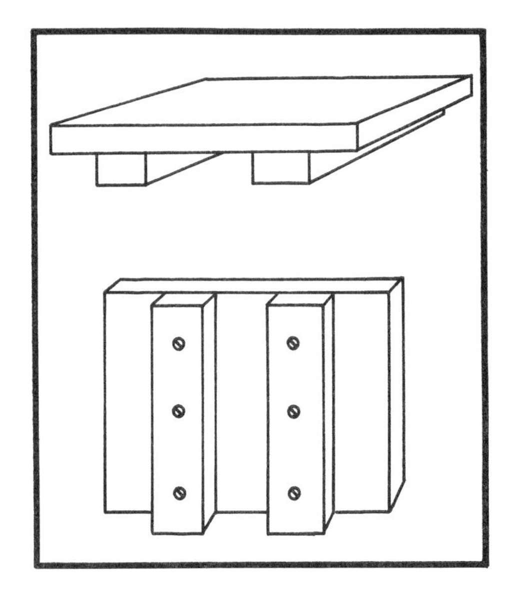

Fig. 4-19. Molding board.

Bottom Board

Bottom board is the board used to support the sand mold until the mold is poured. It is constructed like the molding board, but need not be smooth, only level and stiff. Bottom boards from 10 inches × 16 inches to 18 inches × 30 inch should be made of 1 inch thick lumber with 2 cleats made of 2 × 3 inch stock. Bottom Boards 48 × 30 inch use 1½ inch thick lumber with 3 cleats made of 3 × 3 inch stock. Bottom boards 50 to 80 inches use 2 inch thick lumber with four to five cleats made of 3 × 4 inch stock.

MOLD WEIGHTS

All snap flask work is weighted before molding. In most cases all that is needed is a standard snap weight, which is usually cast by the foundry to suit their own snap sizes. See Fig. 4-20.

The snap flask mold weights are from 1½ to 2 inches thick cast iron and weigh from 35 to 50 pounds for a 12 × 16 inch mold. The rounded cross shaped opening through the weight is to accommodate pouring the metal into the mold. The weights are always set so that the pouring basin is free to easily see and pour, not too close to the weight.

In some cases two or more weights are used per mold, stacked.

Another type of mold weight which I favor for snap and small rigid flask bench work is the sad iron type of weight. They look like

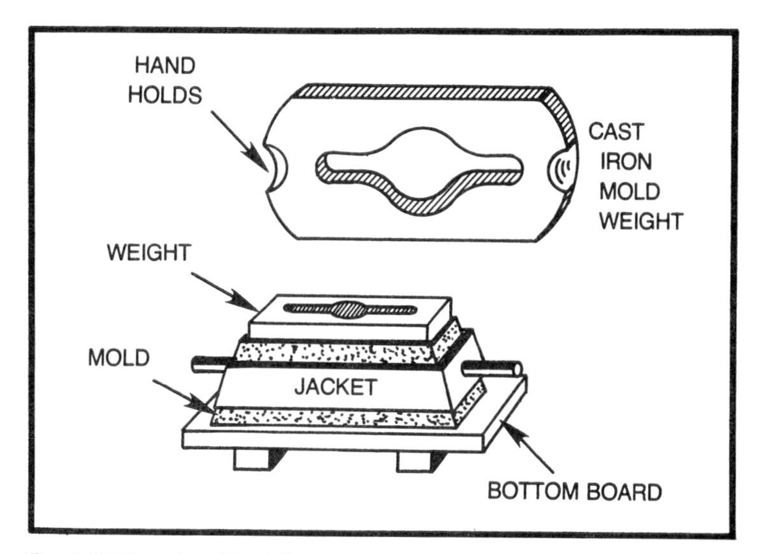

Fig. 4-20. Typical mold weight.

the irons heated on the stove for pressing clothes, only are a good deal heavier. See Fig. 4-21.

Another method used to weight bench molds in rigid full flasks, wood or metal, is to stack the molds three high like steps putting a weight only on the top mold. This practice was quite common in a great many foundries and is still used today. See Fig. 4-22.

The lifting or buoyant force on the cope is the product of the horizontal area of the cavity in the cope, the height of the head of metal above this area and the density of the metal. This lifting force on the cope has to be reckoned with. If the force is greater than the weight of the cope the cope will be lifted by this force when the mold is poured, and run out at the joint resulting in a lost casting plus a mess. In order to determine how much weight has to be placed on the cope, if any, to prevent a lift of the cope we multiply the area of

Fig. 4-21. Sad iron weight.

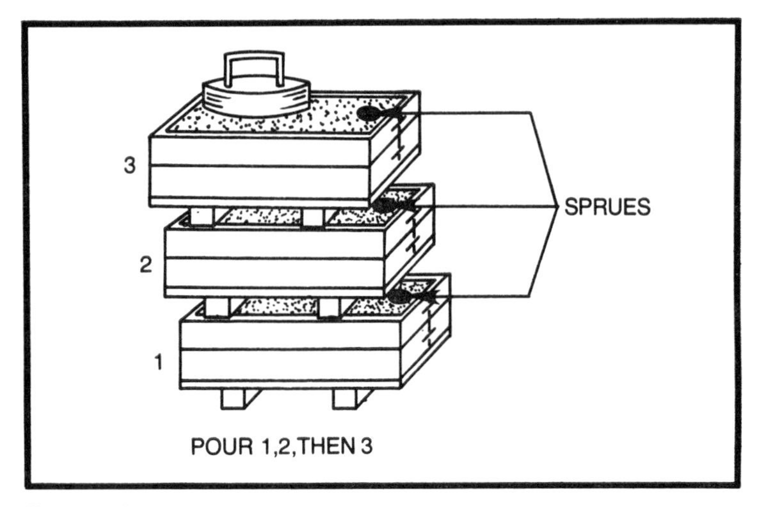

Fig. 4-22. Stacked molds.

the surface of the metal pressing against the cope by the depth of the cope above the casting and the product by 0.26 for iron, 0.30 for brass, and 0.09 aluminum. These figures represent the weight in pounds of 1 cubic inch of iron, brass and aluminum respectively.

So, if we are casting, a flat plate 12 × 12 inches the surface area against the cope would be 144 square inches. If this was molded in a flask which measured 18 × 18 inches with a 5 inch cope and the metal we are casting is red brass, we have 144 × 5 inches the height of the cope, times 0.30 (the weight of 1 cubic inch of brass). We wind up with a lifting force of 216 pounds. One cubic inch of rammed molding sand weighs 0.06 pounds. So with that factor, our cope weight would be 18 × 18 × 5 inches times 0.06 or 97.2 pounds. So, now if we subtract the cope weight from our lifting force 216 pounds minus 97.2 pounds we are still short by about 119 pounds. This would be the weight we would have to add to the cope to prevent it from lfting, with no safety factor. With a 20 percent safety factor add 24 more pounds. Se we should put 150 pounds on the cope.

Chapter 5
Patterns

A pattern is a shaped form of wood or metal around which sand is packed in the mold. When the pattern is removed the resulting cavity is the exact shape of the object to be cast.

The pattern must be designed to be easily removed without damage to the mold. It must be accurately dimensioned and durable enough for the use intended. Either one time use or production runs.

PATTERN MAKING

Each different item we wish to cast presents unique problems and requirements. In a large foundry there is a close relationship between the pattern maker and the molder. Each is aware of the capabilities and limitations of his own field.

Throughout the industry, pattern making is a field and an art of it's own. The pattern maker is not a molder nor the molder a pattern maker. This is not to imply that the pattern maker cannot make a simple mold or the molder make a simple pattern but each may soon reach a point in the other's field beyond his own skill and experience.

In the hobby or one man shop, however, pattern and mold making are so closely interrelated as to become almost one continuous operation. This chapter will acquaint you with some of the various types of patterns and their requirements.

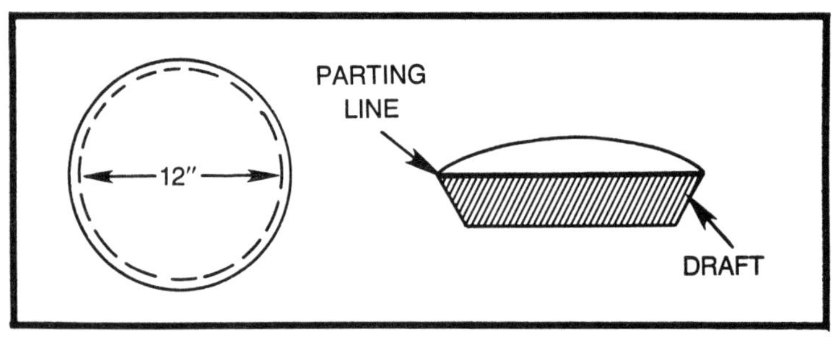

Fig. 5-1. Disc pattern.

Draft

In order to illustrate some of the important pattern characteristics we will use as an example a simple disc pattern. The object we want to cast is 12 inches in diameter and 1 inch thick. The edge of the disc is tapered from the top face to the bottom face. See Fig. 5-1. This taper is known as the pattern draft. This draft is necessary in order that the pattern can be removed easily from the mold causing no damage to the sand. Pattern draft is defined as the taper on vertical elements in a pattern which allows easy withdrawal of the pattern from the mold. The amount of draft required will vary with the depth of the pattern. The general rule is ⅛ inch taper to the foot which comes out to about 1 degree and on shallow patterns such as our disc 1/16 inch taper or 0.5 degree is sufficient.

Shrinkage

Now back to our simple disc pattern. If we wish the casting to come out as cast to the dimensions we show of 12 inches in diameter 1 inch thick, we must make the pattern larger and thicker than 12 inches × 1 inch to compensate for the amount that the metal will shrink when going from a liquid to a solid. This is called pattern shrinkage. This varies with each type of metal and the shape of the casting.

The added dimensions are incorporated into the pattern by the pattern maker by using what is called shrink rulers. These rulers are made of steel and the shrinkage is compensated for by having been worked proportionately over its length. Thus a 3/16 inch shrink rule 12 inches long will be actually 12 3/16 inches long, but for all appearances will look like a standard rule. But, when layed out against a standard ruler it will project 3/16 inch past the standard ruler. These rulers come in a large variety of shrinks. Generally the shrinkage allowance for brass is 3/16 inch per foot, ⅛ inch per foot for cast iron, ¼ inch per foot for aluminum and ¼ inch for steel. This would hold true for most small to medium work, for larger work the shrinkage allowance is less, in some cases 50 percent less. Where a small steel casting in steel would require ¼ inch per foot shrinkage allowance, a very large steel casting might require only ⅛ inch per foot shrinkage allowance. So, from this we see that if we wish to cast a bar in brass 1 foot long we must make the pattern 1 foot and 3/16 inches long to start with.

Machining Allowance

Now the plot thickens, say the disc we want in brass requires that the outer diameter of the casting is to be machined (the 12 inch dimension is a machined dimension). We must then allow for machining to our 12 inch dimension. This allowance must be in addition to the shrinkage and draft allowance, taken at the short side of the pattern or smallest diameter. See Fig. 5-2.

We must have a pattern dimension of 12 3/16 inches the 3/16 inches to allow for shrinkage plus 1/16 inch for metal to come off. So we need an actual diameter on the small end of our pattern of 12⅜ inches.

If we dimension our layout as 12 and 1/16 inches (the 1/16 inch for machining) and we use a 3/16 inch shrink ruler to measure this dimension, then when you build the pattern it will come out fine. Or, make your pattern layout read 12 inches in diameter taking the 12 inch dimension off of a ⅜ inch shrink ruler.

Approximate finish allowances including the draft are, brass 1/16 inch, aluminum ⅛ inch, cast iron ⅛ inch, cast steel ¼ inch.

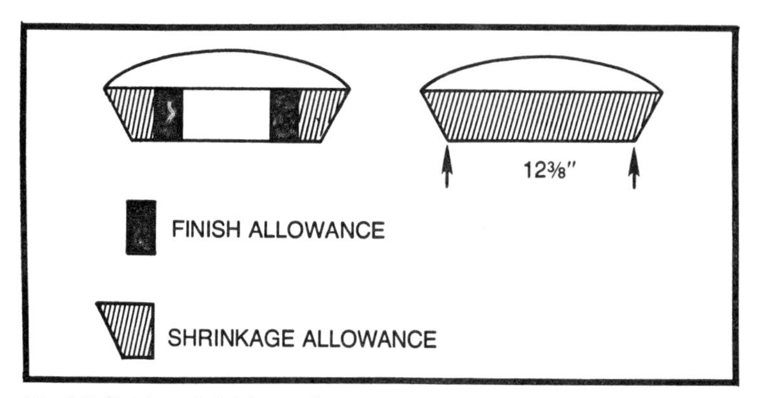

Fig. 5-2. Finish and shrinkage allowance.

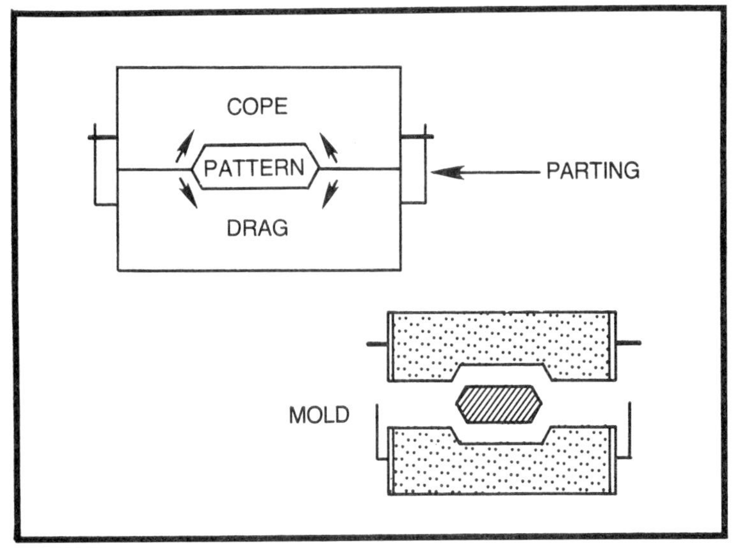

Fig. 5-3. Parting line.

On a blue print given to the pattern maker all finishes should be noted (turned, ground etc.) he will then know from experience how much to allow and how much shrinkage to add to the pattern.

Production Pattern

In the case of our disc, say we are only going to make a casting from the pattern now and again one at a time, we dimension as above. The pattern is called a production pattern, one from which the actual castings are produced.

Master Pattern

Now suppose we want to make one or more production patterns out of cast aluminum from which we intend to make production aluminum castings. In this case we need a wood pattern from which to cast our production pattern. If we wanted as our finished or end product a cast aluminum disc, we would have to make our wood pattern with a double aluminum shrink rule or ½ inch per foot shrinkage. As we are going to take ¼ inch shrinkage in going from our wood pattern to our cast pattern and another ¼ inch to our end product, these rules are called double shrink rulers. If we were going from a wood pattern to an aluminum production pattern to a brass casting as an end product the shrinkage allowance on your wood pattern would have to be ¼ × 3/16 or 12/16 inch plus finish etc., if any. This type of pattern (the wood) is called a master pattern, a pattern from which the production pattern or patterns are made.

Parting Line

On our simple disc pattern of Fig. 5-1, we note the upper face of the pattern is designated as the parting line or parting face. By this we mean a line or the plane of a pattern corresponding to the point of separation between the cope and drag portions of a sand mold. The parting may be irregular or a plane, as the mold must be opened, the pattern removed and then closed for pouring without damage to the sand. The parting line must be located where this can be accomplished. The portion of the pattern in the cope must be drafted so the cope can be removed and the same of the drag. See Fig. 5-3.

Any vertical portion of the pattern in either the cope or drag portion must be drafted or tapered as shown in Fig. 5-4. The junction or change of draft angle indicates the proper position of the parting line.

Back Draft

If the pattern were shaped as in Fig. 5-5 and the mold parted at its upper face the back draft would prevent its removal without damage to the mold.

Back Draft is a reverse taper which prevents removal of a pattern from the mold.

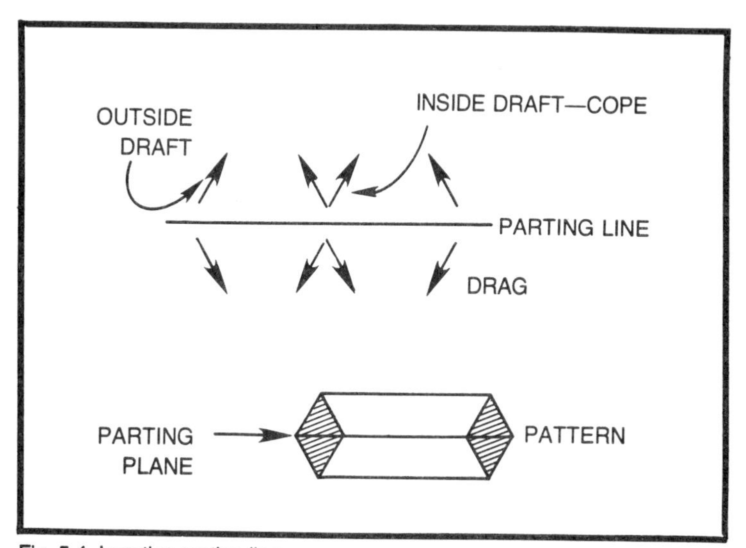

Fig. 5-4. Locating parting line.

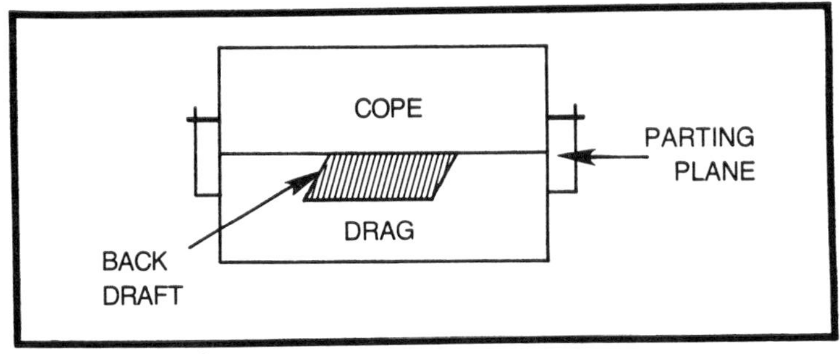

Fig. 5-5. Back draft.

GATED PATTERN

The gate is the channel or channels in a sand mold through which the molten metal enters the cavity left by the pattern. This channel can be made in two ways, one by cutting the channel or channels with a gate cutter, or by the pattern having a projection attached to the pattern which will form this gate or gates during the process of ramming up the mold. See Fig. 5-6.

Set Gate Pattern

If a pattern is made for a gate but not attached to a pattern but only placed against it while making the mold, this pattern is called a set gate pattern.

SPLIT PATTERN

This is a pattern that is made in two halves split along the parting line. The two halves are held in register by pins called pattern dowels. The pattern is split to facilitate molding. See Fig. 5-7.

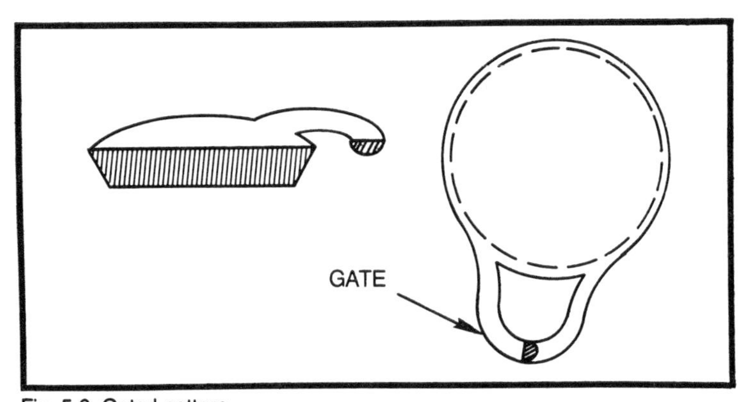

Fig. 5-6. Gated pattern.

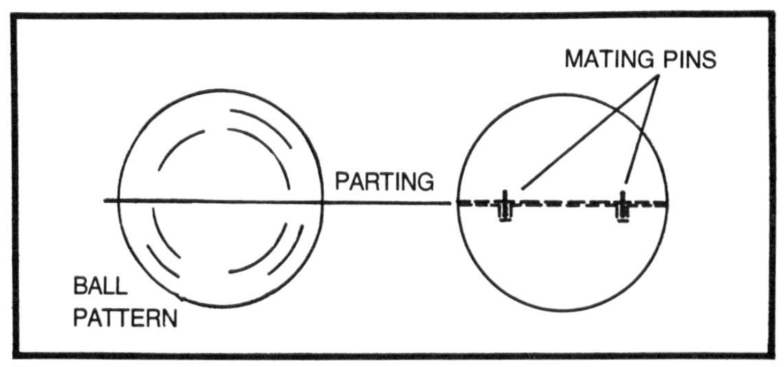

Fig. 5-7. Split pattern.

The dowels hold the two halves of the pattern together in close accurate register, but at the same time are free enough that the two halves can be separated easily for molding. Like the pins and guides of the flask.

The dowels are usually installed off center in such a manner that the pattern can only be put together correctly. See Fig. 5-8.

MEDIUM PATTERN

A pattern that is used only occasionally or for casting a one time piece is usually constructed as cheaply as possible. If it is a split pattern wood dowels are used for pins and fit into holes drilled into the matching half. This type of pattern is called a "Medium Pattern".

A core is a preformed baked sand or green sand aggregate inserted in a mold to shape the interior part of a casting which cannot be shaped by the pattern.

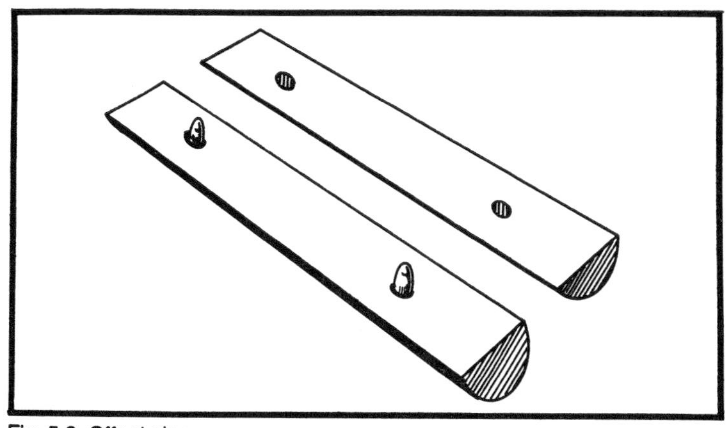
Fig. 5-8. Offset pins.

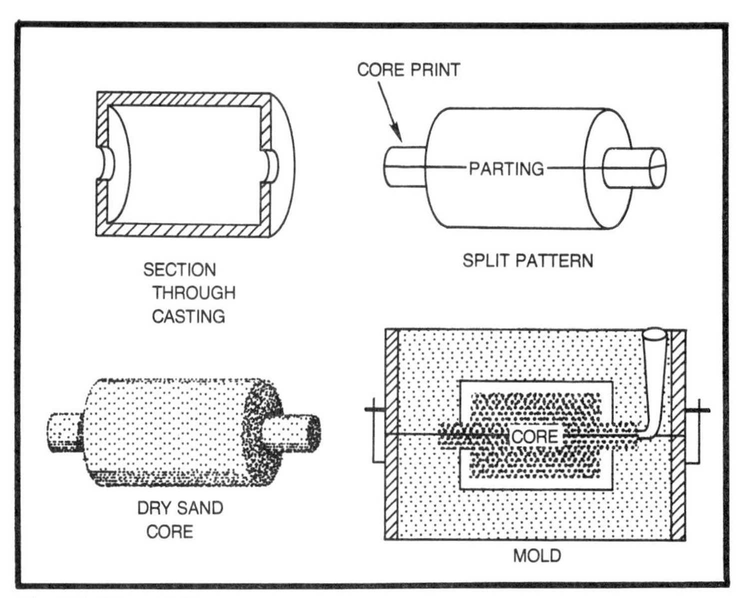

Fig. 5-9. Core prints.

When a pattern requires a core a projection must be made on the pattern, this projection forms an impression in the sand of the mold in which to locate the core and hold it during the casting. These projections are called core prints and are part of the pattern. See Fig. 5-9.

Sometimes it is possible to make a pattern in such a way that a core will remain in the sand when the pattern is removed. The pattern for a simple shoring washer, illustrated in Fig. 5-10 is made in this way.

Mounted Pattern

When a pattern is mounted to a board to facilitate molding, it is called a mounted pattern. In this case the mount has on each end guides which match up with the flask used to make the mold. The plate is placed between the cope and drag flask, the drag rammed and rolled over. The cope is now rammed and lifted off. The plate with pattern attached lifted off of the drag half, the mold finished and closed. See Fig. 5-11.

MATCH PLATE

The match plate is the same as the mounted pattern with the exception that when you have part of the casting in the cope and part

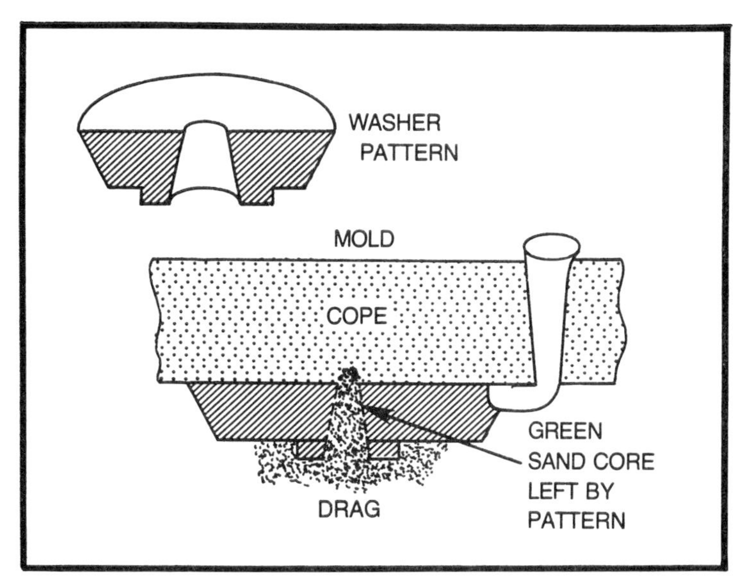

Fig. 5-10. Self-coring pattern.

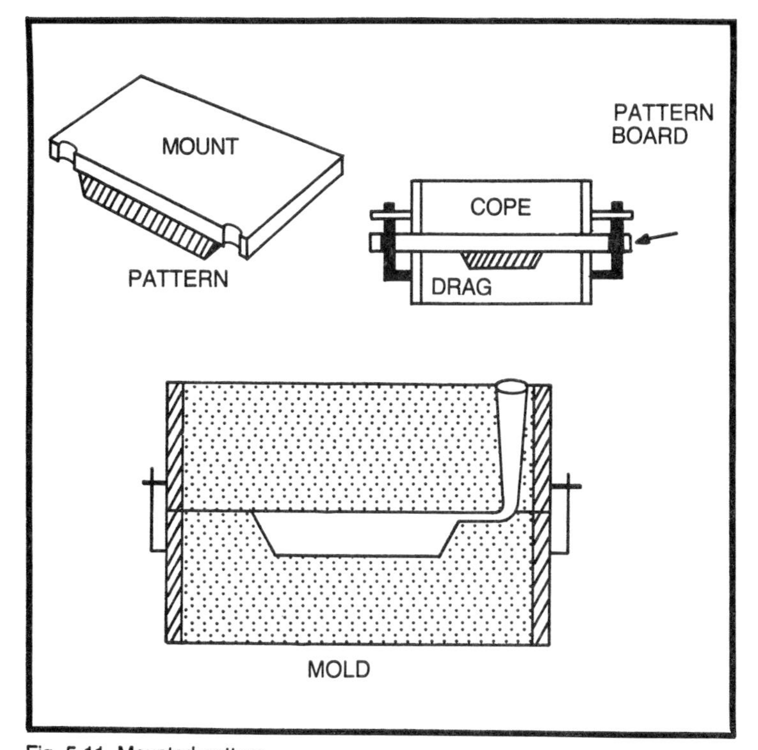

Fig. 5-11. Mounted pattern.

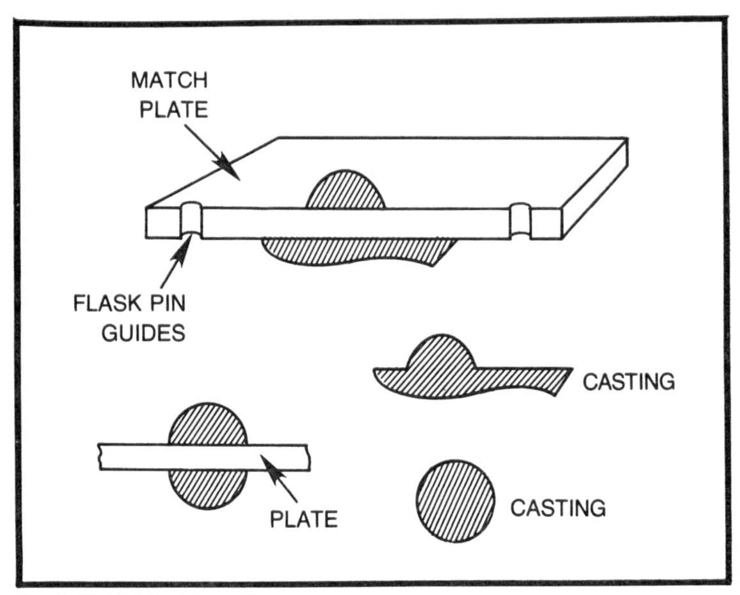

Fig. 5-12. Match plate pattern.

in the drag (split pattern), these parts are attached to the board or plate opposite each other and in the correct location so that when the plate is removed and the mold is closed the cavities in the cope and drag match up correctly. The molding procedure is the same as a one sided mounted plate. See Fig. 5-12.

In most cases all the necessary gating runners, etc., are built right on the plate. The match plate might have only one pattern or a large quantity of small patterns.

Cope and Drag Mounts

In this case you have two separate pattern mounts, one fitted with female guides (for the drag) and one fitted with pins (for the cope). These must match up with the flask used.

The cope half of the pattern is attached to the cope mount and the drag pattern is attached to the drag mount. The cope and drag molds are produced separately and put together for pouring. The usual practice is for one molder to make copes and another to make the drags. Cope and drag mounts are quite common when making large and very large castings where a match plate would be out of the question due to its bulk and weight. Cope and drag mounts are some times called (tubs). See Fig. 5-13.

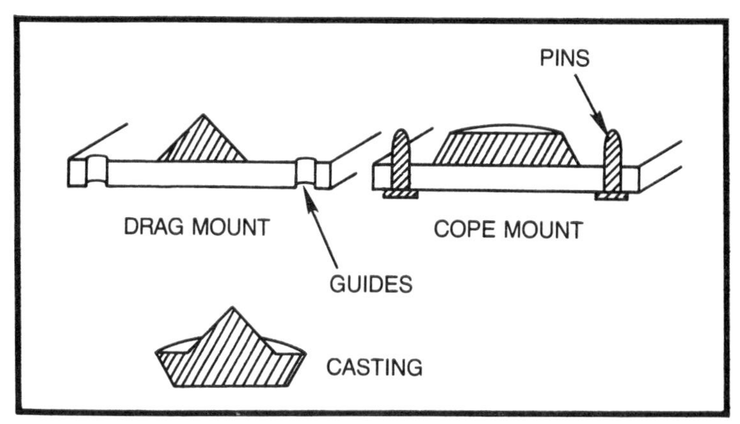

Fig. 5-13. Cope and drag mounts.

Follow Board

A board with a cavity or socket in it which conforms to the form of the pattern and defines the parting surface of the drag. It can be made of wood, plaster or metal. When made of sand it is called a dry sand match.

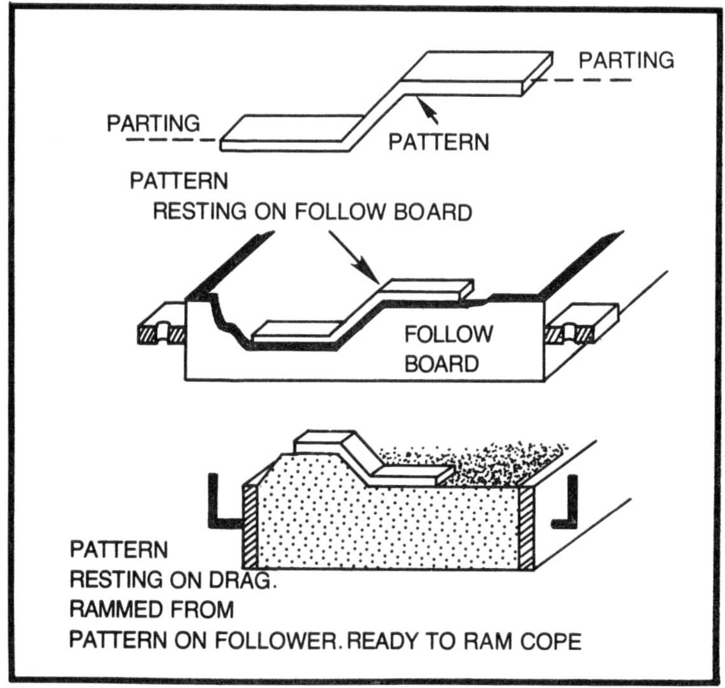

Fig. 5-14. Follow board.

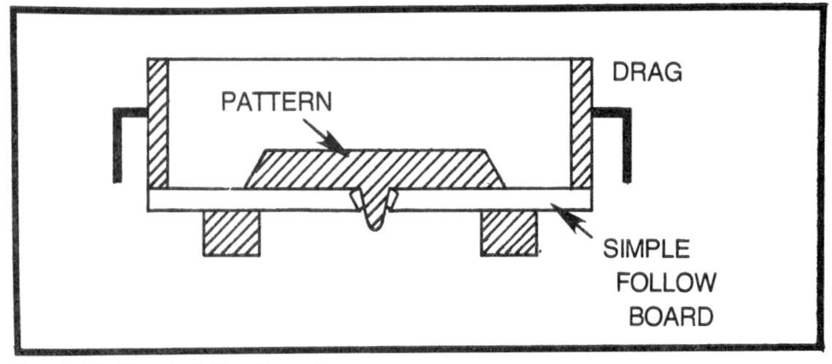

Fig. 5-15. Simple follow board.

The pattern rests in the follow board while making up the drag half of the mold and in doing so, establishes the correct sand parting. The follow board is removed leaving the pattern rammed in the drag up to the parting. The cope then takes the place of the follow board and is rammed in the usual manner. See Fig. 5-14.

A simple follow board might consist of a molding board with a hole in it to allow the pattern to rest firmly on the board while the drag rammed. See Fig. 5-15.

Card of Patterns

When several different loose gated patterns are assembled as a unit to be all molded in the same flask this arrangement is called a card of patterns.

Sweep Pattern

A sweep pattern consists of a board having a profile of the desired mold, which when revolved around a suitable spindle or

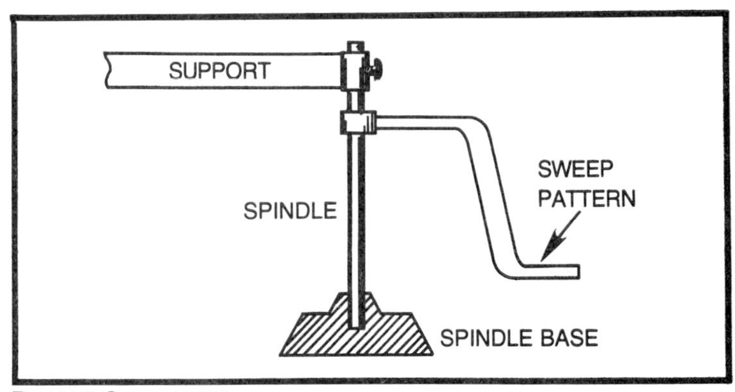

Fig. 5-16. Sweep pattern.

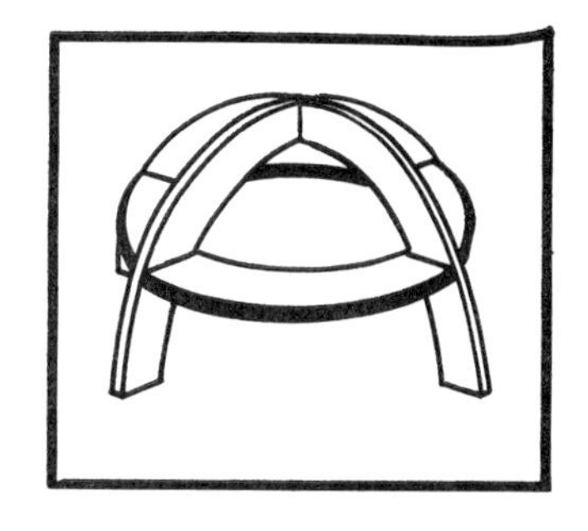

Fig. 5-17. Skeleton pattern.

guide produces that mold. Two are usually required, one to sweep the cope profile and the other the drag profile. See Fig. 5-16.

Skeleton Pattern

This is a frame work of wooden bars which represent the interior and exterior form and the metal thickness of the required casting. This type of pattern is only used for huge castings. See Fig. 5-17.

Expendable Pattern

As in Lost Wax casting the pattern is lost. Expendable patterns for sand casting are styrofoam which is shaped to the desired form with attached styrofoam gates, runners and risers. The styroform pattern is molded with dry clay-free sharp silica sand in a box or steel frame. The pattern is vaporized by the metal poured into the mold, leaving the casting.

WOOD PATTERNS

Wood patterns used for sand casting are given several coats of Orange Shellac to which a pinch of oxalic acid has been added. This gives them a good waterproof smooth hard surface.

The majority of wood patterns are made of white pine (sugar pine) as it is easily worked and when shellacked properly will not warp under ordinary foundry use.

The approximate weight of a casting can be determined by weighing the wood pattern and multiplying by the appropriate factor indicated. Aluminum 8, cast iron 16.7, copper 19.8, brass 19.0, steel 17.0.

A white pine pattern weighing 1 lb, when cast in aluminum will weigh 8 pounds, in brass 19 pounds etc.

Chapter 6
Cores and Core Boxes

A core is a preformed baked sand or green sand aggregate inserted in a mold to shape the interior part of a casting which cannot be shaped by the pattern.

A core box is a wood or metal structure, the cavity of which has the shape of the desired core which is made therein.

A core box, like a pattern is made by the pattern maker. Cores run from extremely simple to extremely complicated. A core could be a simple round cylinder form needed to core a hole through a hub of a wheel or bushing or it could be a very complicated core used to core out the water cooling channels in a cast iron engine block along with the inside of the cylinders.

Dry sand cores are for the most part made of sharp, clay-free, dry silica sand mixed with a binder and baked until cured, the binder cements the sand together. When the mold is poured the core holds together long enough for the metal to solidify, then the binder is finely cooked, from the heat of the casting, until its bonding power is lost or burned out. If the core mix is correct for the job, it can be readily removed from the castings interior by simply pouring it out as burnt core sand.

This characteristic of a core mix is called its collapsibility. The size and pouring temperature of a casting determines how well and how long the core will stay together. A core for a light aluminum casting must collapse much more readily than a core used in steel, because of the different time, weight and heats involved. Core sand,

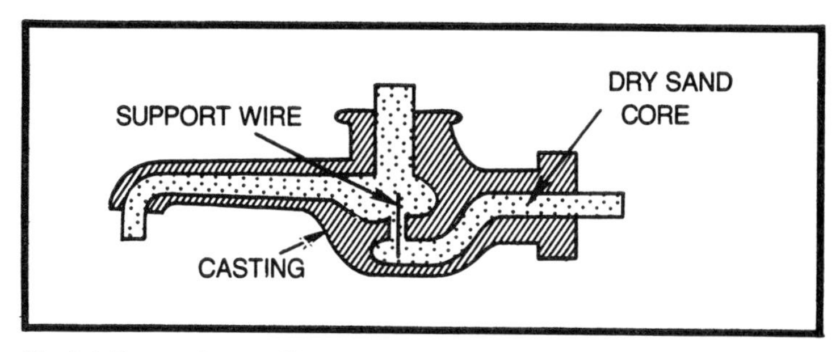

Fig. 6-1. Dry sand core with support wire.

like molding sand, must have the proper permeability for the job intended. The gases generated within the core during pouring must be vented to the outside of the mold preventing gas porosity and a defect known as a core blow. Also, a core must have sufficient hot strength to be handled and used properly. The hot strength refers to its strength while being heated by the casting operation. Because of the shape and size of some ores they must be further strengthened with rods and wires.

A core for a cast bronze water faucet must have a wire in the section which passes through the seat of the valve. Because of the small diameter at this point core sand alone would not have sufficient dry strength to even get the core set in the mold much less pour the castings. See Fig. 6-1.

A long span core for a length of cast iron pipe would require rodding to prevent the core from sagging or bending upward when the mold is poured because of the liquid metal exerting a strong pressure during pouring. See Fig. 6-2.

BINDERS

There are many types of binders to mix with core sand. A binder should be selected on the basis of the characteristics that are most suitable for your particular use.

Some binders require no baking becoming firm at room temperature such as rubber cement, Portland cement and sodium silicate or water glass. In large foundry operations and in some school foundries, sodium silicate is a popular binder as it can be hardened almost instantly by blowing carbon dioxide gas through the mixture.

Oil binders require heating or baking before they develop sufficient strength to withstand the molten metal. Linseed oil is considered one of the strongest binding oils and vegetable, mineral and fish oils are also used.

Sulfite binders also require heating. The most popular of the sulfite binders is a product of the wood pulp industry. It is sold in liquid form under the trade name of Glutrin and in the dry form under the trade name of Goulac.

Furan is the trade name of a chemical binder which is hardened by the use of a catalyst. There are many liquid binders made from starches, cereals and sugars. They are available under a countless number of trade names.

A good binder will have the following properties;

- Strength
- Collapse rapidly when metal starts to shrink.
- Will not distort core during baking.
- Maintain strength during storage time.
- Absorb a minimum of moisture when in the mold or in storage.
- Withstand normal handling.
- Disperse properly and evenly throughout the sand mix.
- Should produce a mixture that can be easily formed.

CORE MIXES

The following is a list of core mixes that I have used for a number of years with good success.

For brass and aluminum

Fine river sand .... 10 parts
Wheat flour .... ¼ parts
Air float sea coal .... 3 parts
Rosin .... 1 part

For small castings

New molding sand .... 15 parts
Sharp sand .... 5 parts
Linseed oil .... 1 part

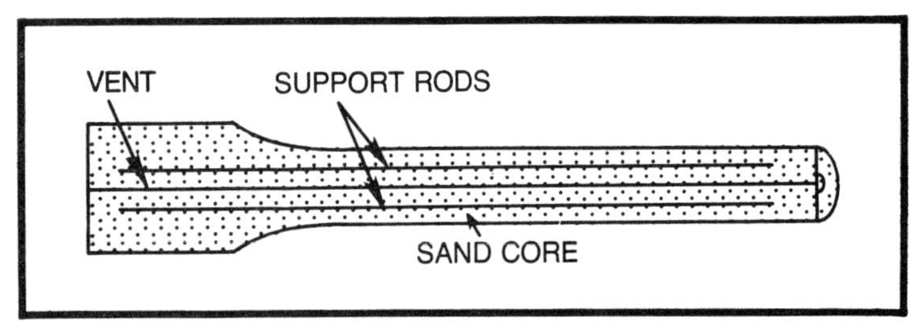

Fig. 6-2. Rodded core.

For aluminum, heavy castings

Sharp sand ....30 parts
New molding sand..10 parts
Wheat flour...2 parts
Temper with molasses water.

For aluminum, medium castings

Sharp sand ....10 parts
New molding sand ....5 parts
Wheat flour ....1 part
Temper with molasses water.

Good mix for long skinny cores

Sharp sand ....8 quarts
Wheat flour...1 quart
Core oil ....⅛ pint

Quick collapse cores for aluminum

Sharp sand ....45 parts
Molding sand ....45 parts
Powdered rosin...2 parts
Wheat flour ....1 part

Mix for small cores

Sharp sand...25 quarts
Molding sand...15 quarts
Linseed oil..1 quart

Brass pump core mix

Silica..32 parts
Fire clay..1 part
Silica flour ....4 parts
Add rosin to desired dry strength.

Small diameter bushing cores

Sharp sand ....80 parts
Corn flour...2 parts
Oil ....1 part

Now that if you are confused about core mixes, don't let it worry you. Foundry A produces the same type and weight of castings as Foundry B. Foundry A has 100 or more different core mixes. They use almost a different core mix for each job. Foundry B doing the same kind of work uses 3 different mixes, one for light work, one for medium work and one for heavy.

Both produce good work. However, I am sure foundry B cost and problems are much less than A. This condition exists throughout the industry. Each and every core room foreman has his own pet mixes that he carries throughout life from shop to shop as does every core maker. For general all around core work in the small or hobby shop, a simple linseed oil bonded mix, this a universal core mix will suffice for 90 percent of the work.

The simplest mix is by volume, 40 parts fine sharp washed and dried silica sand and 1 part of linseed oil.

It is common practice to add a small percentage of water and kerosene. This makes it mix better and strip easier from the core box.

If the moisture is too low in the mix the core will bake out too soft and if the moisture is too high the core will bake out too hard and stick in the box when making the core. Most mixes work best when just enough water is added to make the mix feel damp but not wet. To mix your core mix start out with dry sand, nearly pure silica sand free from clay. Mix the dry ingredients, add the oil and water and finish mixing.

Core sand mixes can be mixed in a muller or paddle type mixer and in small amounts on the bench by hand.

The core is made by ramming the sand into the core box and placing the core on a core plate to bake. See Fig. 6-3.

The box cavity is dusted with parting powder usually made of powdered walnut shells, purchased as a core box dry parting. The box is rammed full of sand using the handle of a raw hide mallet to ram with. The excess is struck off level with the side of a core maker's trowel, the core is then vented with a vent wire, a core plate placed on top and the plate and box rolled over on the bench. The box is rapped on one side and then on the front and back to loosen the core.

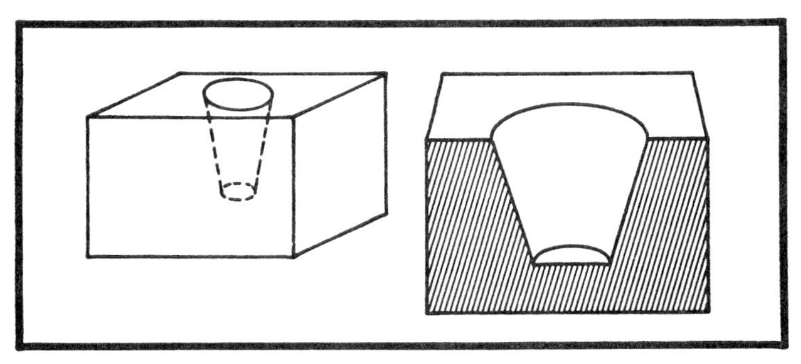

Fig. 6-3. Simple dump core boxes.

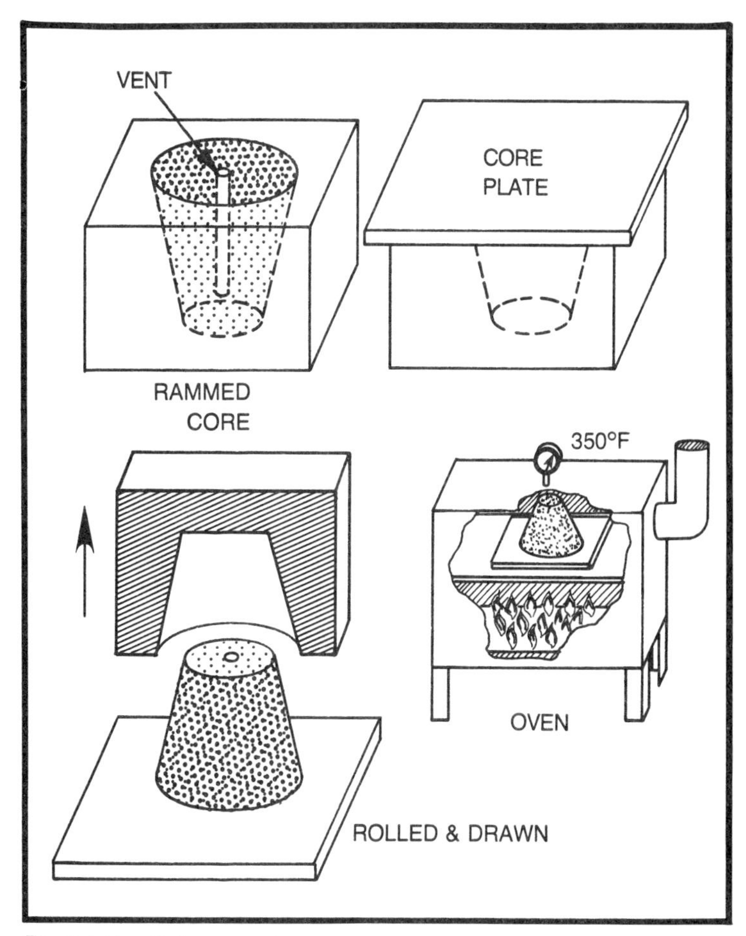

Fig. 6-4. Use of dump core box.

The box is the lifted (drawn) off leaving the core on the plate. The plate is then placed in the oven to be baked or dryed. When finished, it is removed from the oven and allowed to cool. It is then ready to use. See Fig. 6-4.

Stand Up Cores

Stand up cores (cores that can be stood on end to dry) are made in a split core box. The box is pinned together with pattern bushings. The box is held together with a C-clamp, it is rammed up using a dowel of suitable diameter. When rammed, it is then vented with a vent wire, the clamp removed and the box placed on a core plate. The half of the box away from the core maker is removed and the

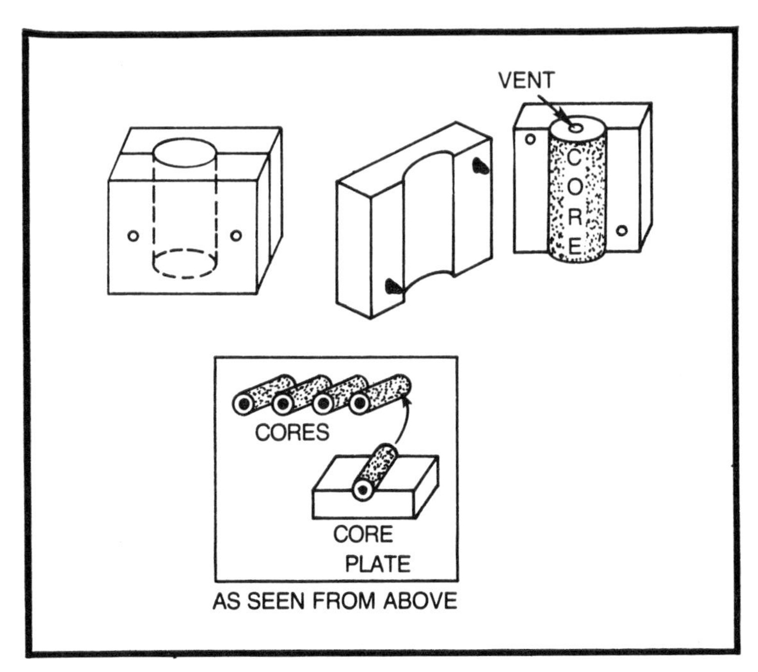

Fig. 6-5. Stand up core box.

core slid to the far side of the plate, the remaining half of the box is removed leaving the core standing free. This is continued until the core plate is full, or you have sufficient cores. The plate is then ready for the core oven. See Fig. 6-5.

Stand up cores are often made in gang core boxes. See Fig. 6-6.

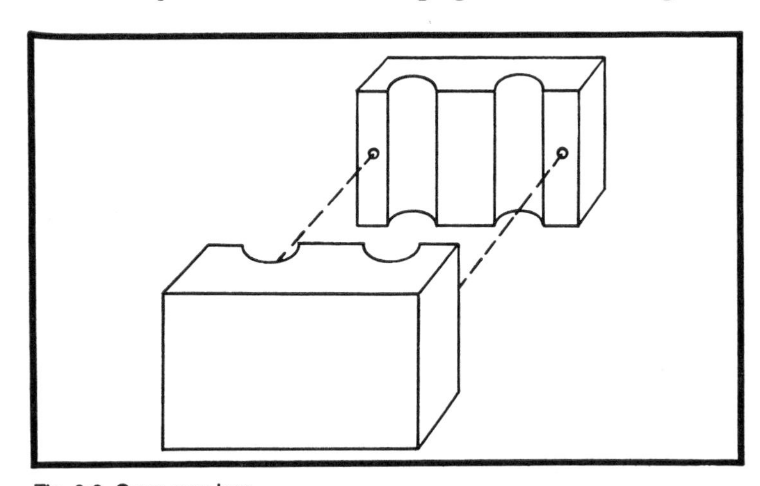

Fig. 6-6. Gang core box.

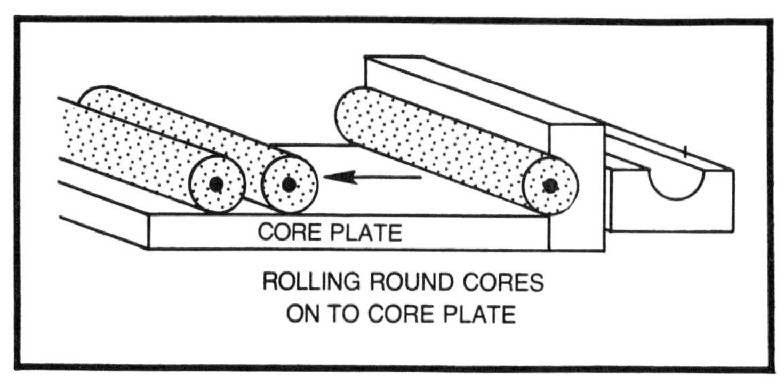

Fig. 6-7. Use of core plate.

Stock or round cores when too tall to stand on end are made in a split box and rolled on to a core plate for drying. See Fig. 6-7.

Pasted Cores

Cores can be made in halves and after they are dried, glued together to make the complete core. If the core is symmetrical, a half box is all that is needed. See Fig. 6-8.

After the two halves of the core are dried, a vent is scratched along the center line of one half and the sections are glued together with core paste. The seam is then mudded with a material known as core daubing. Both core paste and daubing can be purchased or made. For home made core paste use enough wheat flour dissolved in cold water to produce a creamy consistency.

For a good core daubing mix fine graphite with molasses water. (One part molasses to ten parts water.) Enough graphite is added to the molasses water to make a stiff mud.

Another good daubing mix is graphite and linseed oil mixed to a stiff mud.

After a core has been pasted and daubed, it is a good idea to return the core to the core oven for a short period to dry the paste and daubing.

A Three Part Core Box

This consists of a top, front and back section. The box is assembled and placed top down and clamped. You ram it up, put on a core plate, roll it over, rap it. Remove the top (which forms the negative section in top of the core) then remove the main box from the core as you would any simple split box. See Fig. 6-9.

Loose Piece Box

This is a box which contains a loose piece (or drawback) to form a particular shape to the core. The loose piece is placed in position in the box, the core rammed, rolled over on a plate, rapped and lifted off. The loose piece remains with the core on the plate and is drawn back to remove it. See Fig. 6-10.

The core box may have one loose piece or many. Sometimes a loose piece may have a loose piece of its own.

Swept Core

A core can be produced by a suitable shaped sweep (or stickel) and guide rails. The guide rails are clamped to a core plate. Core sand is heaped in between the rails and swept to shape by the sweep form. When finished the guide frame is removed and the core dried. The swept core might have several different diameters and may be curved etc. The shape dictating the frame construction and the number and shape of the strikes. Flat cover or slab cores are more

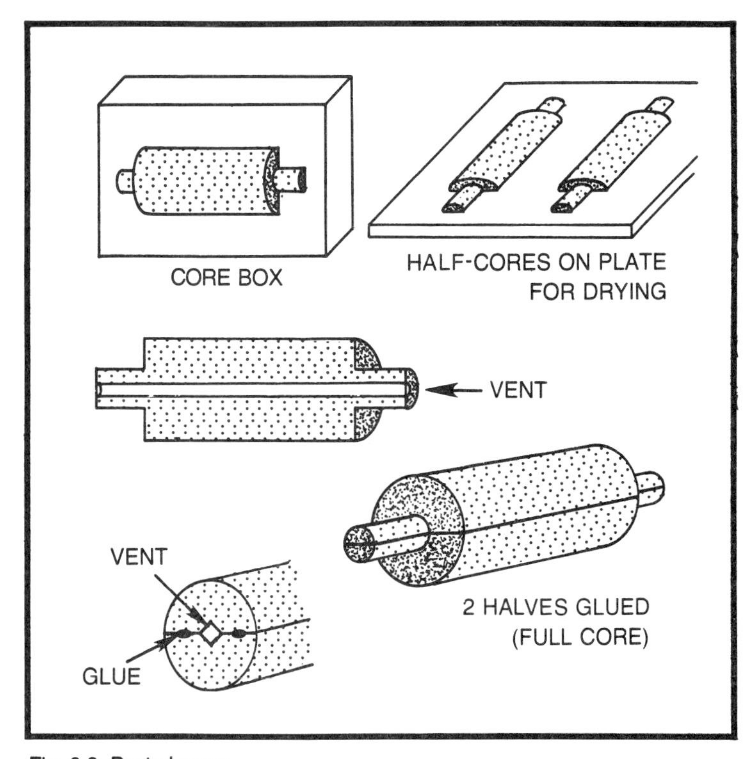

Fig. 6-8. Pasted cores.

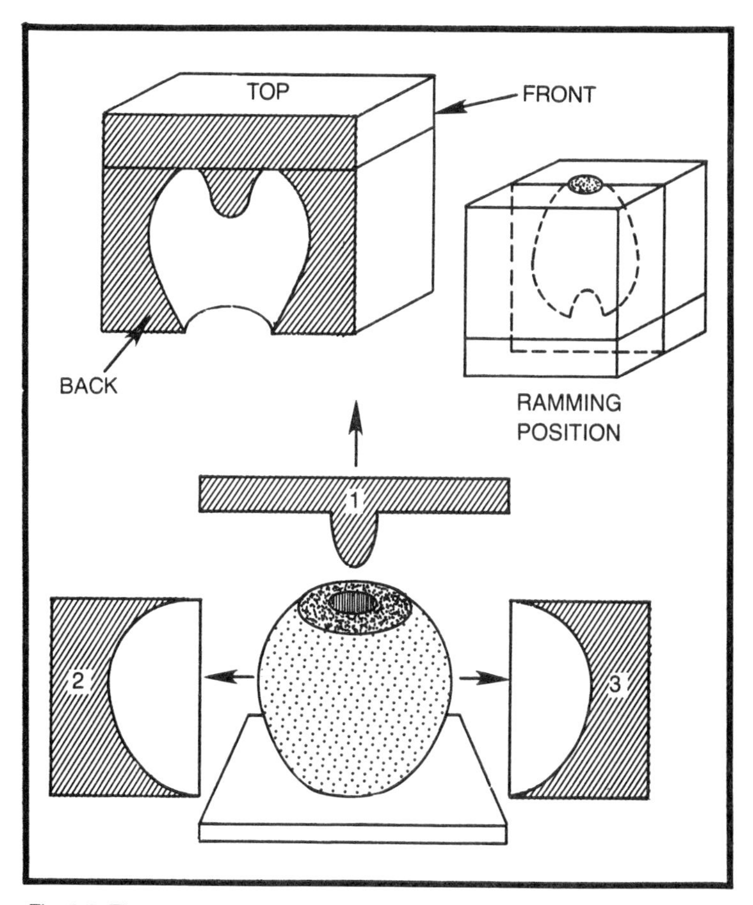

Fig. 6-9. Three part core box.

often than not swept up on a core plate with a simple frame and flat strike off bar. See Fig. 6-11.

CORE DRIERS

A core drier is a support used to keep cores in shape during the baking or drying. Most core driers are made of cast aluminum. The core drier is similar to a half core box but fits the core more loosely than the box and only sufficiently to prevent the core from collapsing or sagging before and during drying. The core is rammed in the usual manner, the side of the core box that presents the face of the core that will rest in the drier is removed. The drier replaces this half box. The box and drier are rolled over, the box removed, leaving the core resting in the drier ready to be dried. See Fig. 6-12.

Green Sand Drier

In this case we simply make a bed of riddled molding sand to support the core during drying. The core box is rammed, one half of the box removed and replaced by a tapered wooden frame. Dry parting is shaken over the exposed half of the green, unbaked core and the rame is filled to the top with lightly riddled tempered fine molding sand and struck off level with the top of the frame, a core plate is then placed on the frame and the whole assembly rolled over. The remaining half of the core box is removed, then the wooden frame. This leaves the core supported in a bed of molding sand for drying. When dry, the green sand is simply brushed off of the baked core with a molder's brush. See Fig. 6-13.

BALANCE CORE

This is when the core is supported on one end only and the other unsupported end extends a good way into the mold cavity. In order to prevent the core from being lifted up by the metal coming into the mold we must extend both the length and diameter of the core section held by the single print. In other words balance the core. See Fig. 6-14.

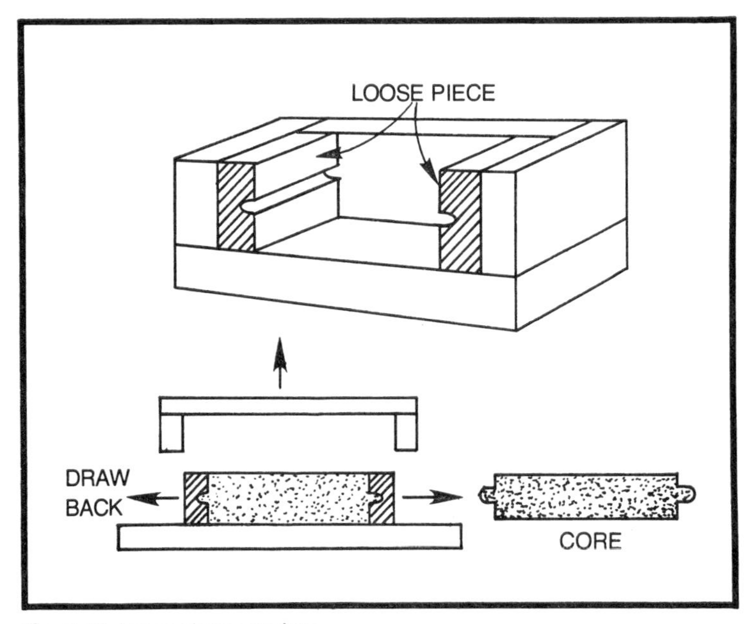

Fig. 6-10. Loose piece core box.

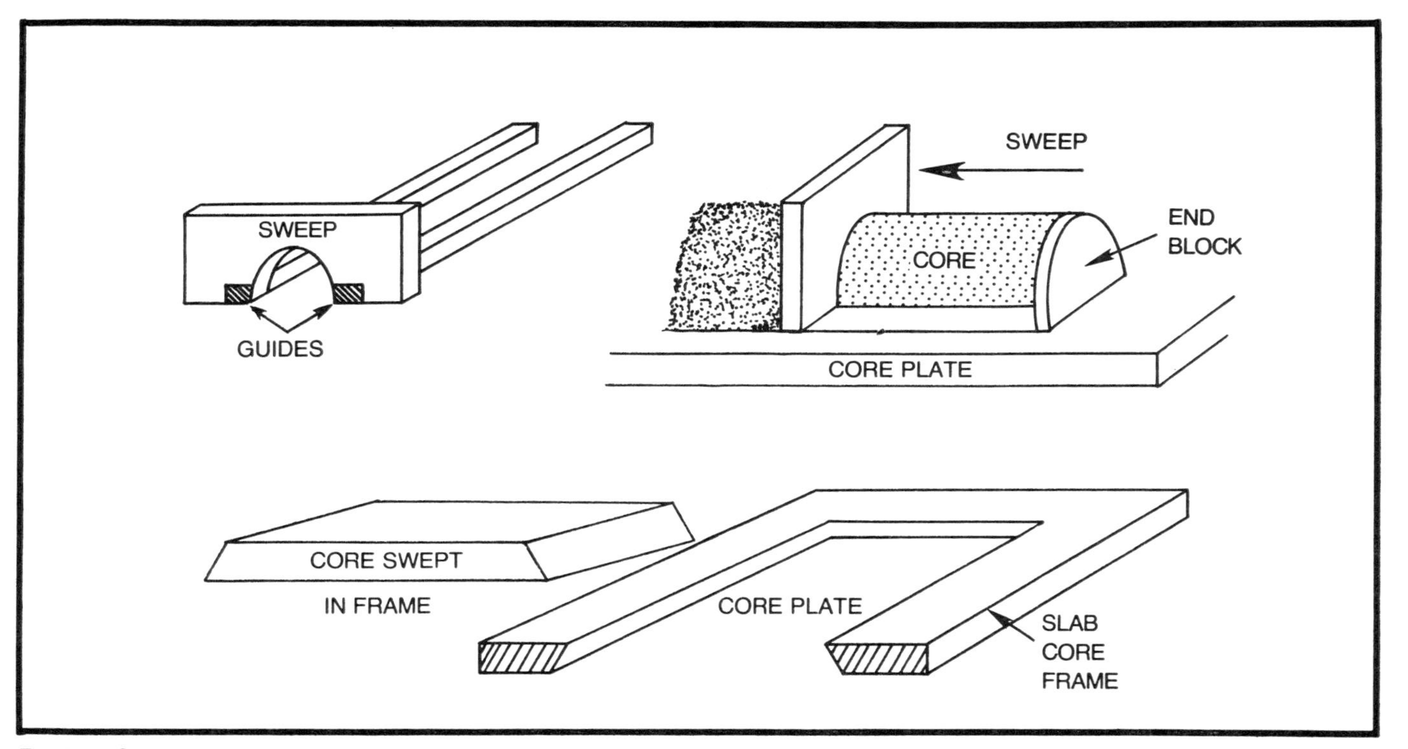

Fig. 6-11. Swept core.

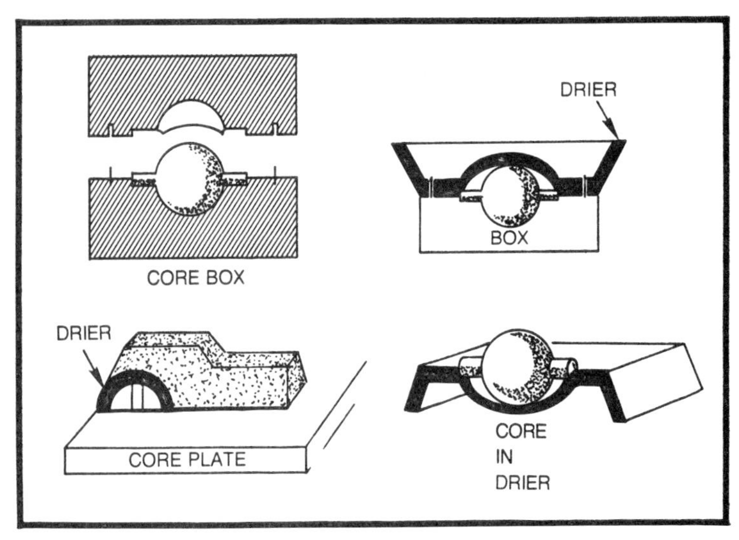

Fig. 6-12. Core driers.

DUMBBELL CORES

To prevent having to make a balanced core which not only takes up room in the flask, but increases core cost, a better solution is to use one core to produce two castings (a core to produce two castings having a common print). See Fig. 6-15.

CHAPLETS

Chaplets consist of metallic supports or spacers used in a mold to maintain cores, which are not self supporting, in their correct

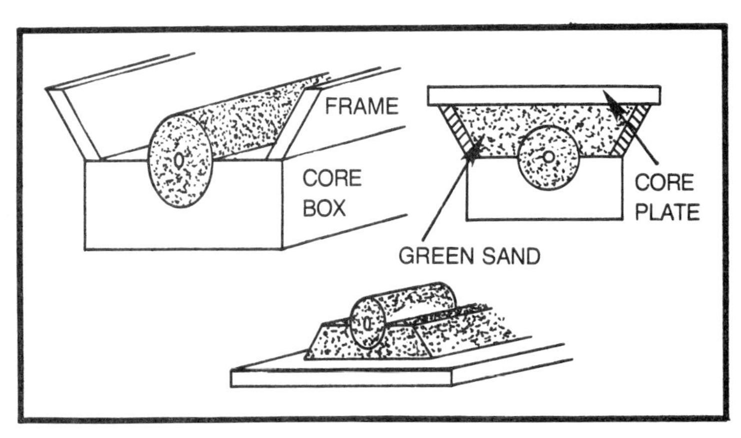

Fig. 6-13. Green sand drier.

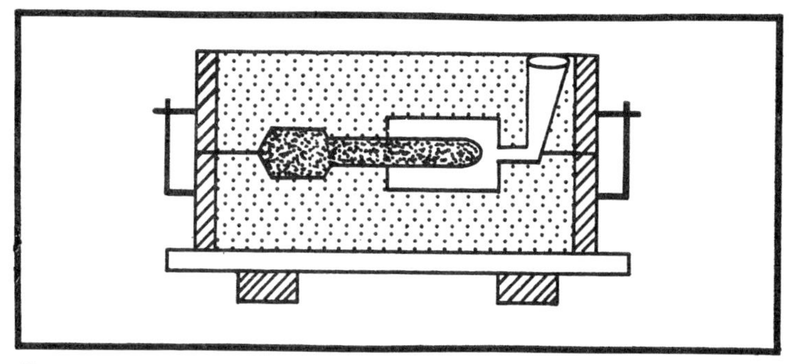

Fig. 6-14. Balance core.

position during the casting process. They are not required when a pattern has a core print or prints which will serve the same purpose. Chaplets are purchased in a wide variety of sizes and shapes to fit just about any need or condition. The three most common types are the stem chaplet, motor chaplet and the radiator chaplet.

Motor chaplets and stem chaplets are placed or set in the mold after the mold is made. Radiator chaplets are rammed up in the mold during the molding operation.

On many gas stove burners or hot water heater burners and many types of gas furnace burners, you will note that there are a series of projections or little bumps on the casting face both on the cope and drag. They are more noticeable on the bottom face. These bumps or knots are made by radiator chaplets used to hold the core centrally located when these burners are cast. The pattern is drilled

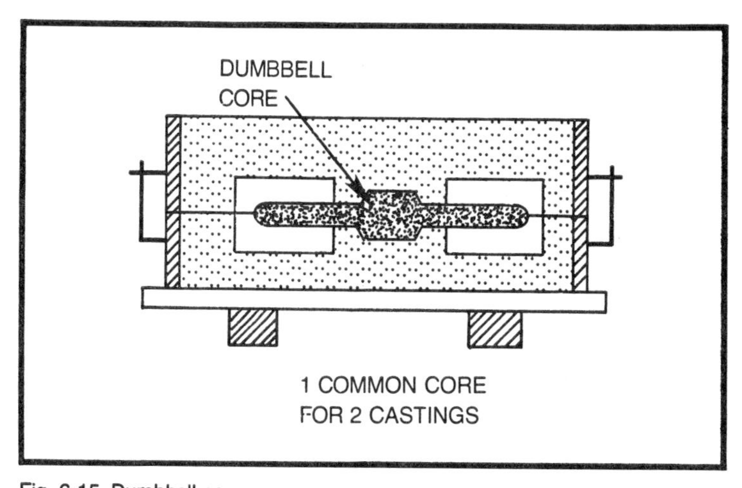

Fig. 6-15. Dumbbell core.

wherever a chaplet is needed. These drilled holes are slightly larger in diameter than the stem diameter. The chaplet has a knot back from the business end of its stem that allows the chaplet to only stick into the hole in the pattern as far as this knot. The distance from the knot to the end of the stem represents the metal thickness of the casting. The end of the chaplet that is held in the molding sand has a plate shaped head to give it a good purchase in the rammed molding sand which prevents it from moving or being pushed back by the force of the core as it tries to float when the mold is being poured.

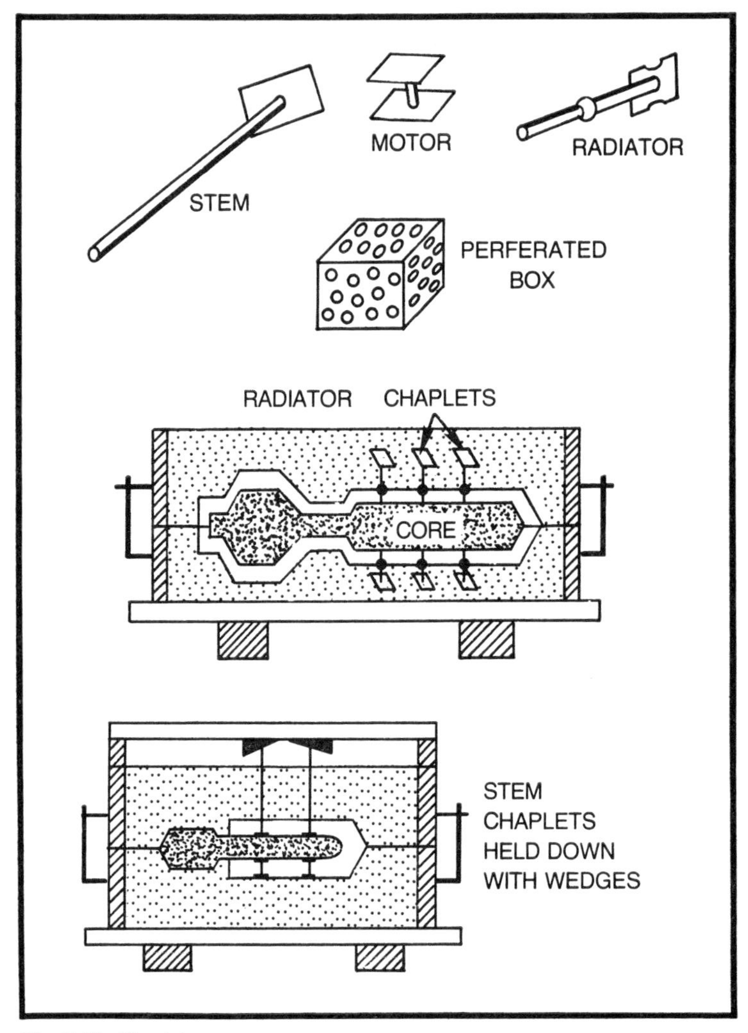

Fig. 6-16. Chaplets.

The molder sets the stem chaplets into the holes in the drag half of the pattern, rams the drag, rolls the flask over and sets the cope chaplets, rams and finishes the cope. When the pattern is removed you have the stems of the chaplets sticking through the molding sand the exact distance (Metal thickness between core and mold cavity) required. The core is set on the drag chaplets and the chaplets in the cope come down and clamp the core between cope and drag chaplets when the mold is closed.

When the casting is poured, the projecting chaplets hold the core in place until the metal starts to solidify, then these projections fuse or weld themselves into the metal that surrounds them.

A leaking hollow casting, sometimes results when using radiator chaplets. The problem is because the casting was not poured hot enough to properly fuse the chaplet, or a rusty or dirty chaplet was used by the molder. After the casting is shaken out, the stem and head that was in the sand is broken off. The chaplet is provided with a break off notch. See Fig. 6-16.

Chaplets are seldom if ever used in non-ferrous casting because the pouring temperature is not high enough to melt and fuse the chaplet.

COVER CORE

A core which is set in place during the ramming of a mold to cover and complete a cavity partly formed by the withdrawal of a loose piece on the pattern.

RAM UP CORE

A ram up core is a core that is set against the pattern or in a locator (slot etc.) in the pattern, the mold rammed and when the pattern is drawn the core remains in the mold. See Fig. 6-17.

As you can see by now there is an endless variety of types, kinds, and uses of cores. New uses and kinds are continuously coming up as new problems present themselves. Therefore we have covered only the most common ones and their uses.

CORE WASHES

Cores are sometimes coated with a refractory wash to increase the cores refractiveness and to produce a smoother metal surface in the cored cavity of the casting. These materials are called washes. They can be purchased in a variety of types and refractive strength. A common home made wash is graphite and molasses water mixed to a nice paint consistency.

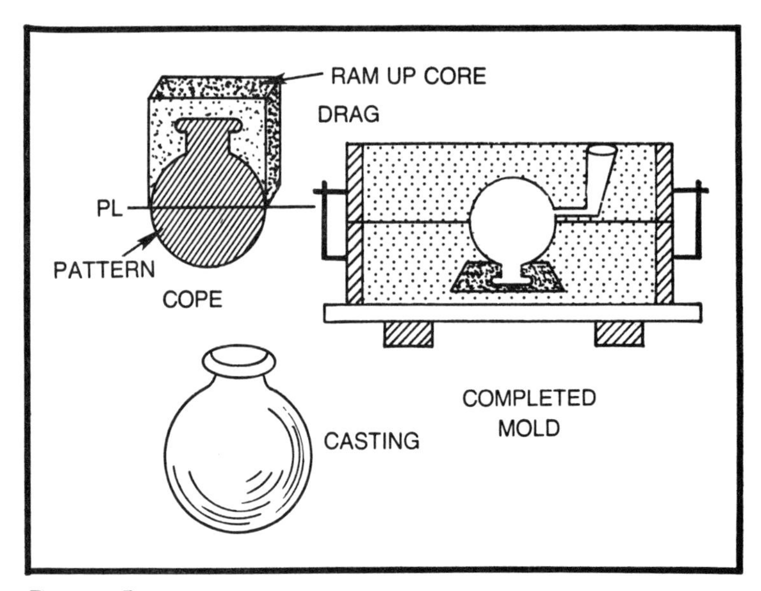

Fig. 6-17. Ram up core.

CORE PLATES

Core plates as you have no doubt guessed by now are flat plates used to bake the ores on in the core oven. They can be made of cast aluminum which has been normalized and machined to a true flat surface. Composition plates can be purchased which are called Transite, a composition asbestos mill board manufactured by Johns Manville Co., these can be had in a range of sizes and thicknesses from any foundry supply house.

VENT WAX

These are tapers of various diameters made of a wax and rosin combination. The vent wax is rammed up in the proper position in a core, and when the core is baked, the wax melts leaving a vent channel. This wax is sold in spools.

CORE BAKING AND CORE OVENS

The cores are baked in order to set the binder. The usual temperature range for oil bonded cores is from 300 to 450 degrees Fahrenheit. The time required varies with the bulk of the core. A large core might take several days to bake or a small core might bake out in an hour or less. When an oil core is completely baked the outside is a rich dark brown not black or burned. The core must be cured completely through with no soft centers. Only experience and trial and error will teach you how to bake cores. One thing for sure, if they are green they can be baked longer until done, but if you burn them up too bad. Start over.

It is common practice to hollow out large cores to decrease the baking time and assure an even bake throughout.

Another factor which relates to the time and temperature required to properly dry a core, is the type and amount of binder used. Oil binders require hotter and quicker baking than rosin, flour and goulac binders.

The core oven, which is usually a gas fired oven with temperature controls is equipped with shelves on which to set the core plates and cores for baking. Some have drawers like a file cabinet which pull out to load and unload. When the drawer is pulled completely out, its back closes off the opening in the oven and prevents heat loss. This type of oven is also made with semi circular shelves which swing out.

The core oven can consist of a square or rectangular brick oven with doors. The bottom of the oven is floor level. The ores are placed on racks which, when full, are rolled into the oven, the oven closed and the cores baked. It is common practice to load the ovens one day with the cores required for the next day and dry them over night. The lead time required for cores to the molder will vary from shop to shop.

It stands to reason the molder cannot be given a job requiring cores unless the cores are ready and on hand in the correct amount.

You can bake cores in your wife's kitchen oven. There are gangs of different kinds of mechanized machinery which can be purchased to make cores, completely automated to semi-automated, blowers, extruders, shell machines etc., but we are not into that.

Chapter 7

Bench Molding

A very important piece of equipment for molding is of course a husky molders bench. A suggested design is shown in Fig. 7-1. The three quarter inch plywood board is the working area, the two by four runners are for rolling over the flasks as described further along in this chapter. The spacing between the runners may vary somewhat depending on the size of the flasks you will be working with. The runners should overhang the edge of the bench far enough to permit the inverted flasks to be placed underneath for the roll over. Four to six inches should be enough. The ends of the runners protruding from the table are rounded off as shown, this makes the roll over possible.

The whole assembly should be supported by 2 by 4 inch lumber. Some benches have small cast iron wheels on the two back legs to permit moving along a windrow, putting the molds behind you and advancing the bench as the sand is used up. See Fig. 7-2.

MOLDING PROCEDURES

In order to produce a sand mold on the bench or floor you have to carry out a number of procedures in a specified order, none more or less important than the other. It is not enough to say ram the mold, cut the gate, roll over the drag, cope out the pattern. You have to know exactly how all operations are carried out and in what order and more important why each is done the way it is done.

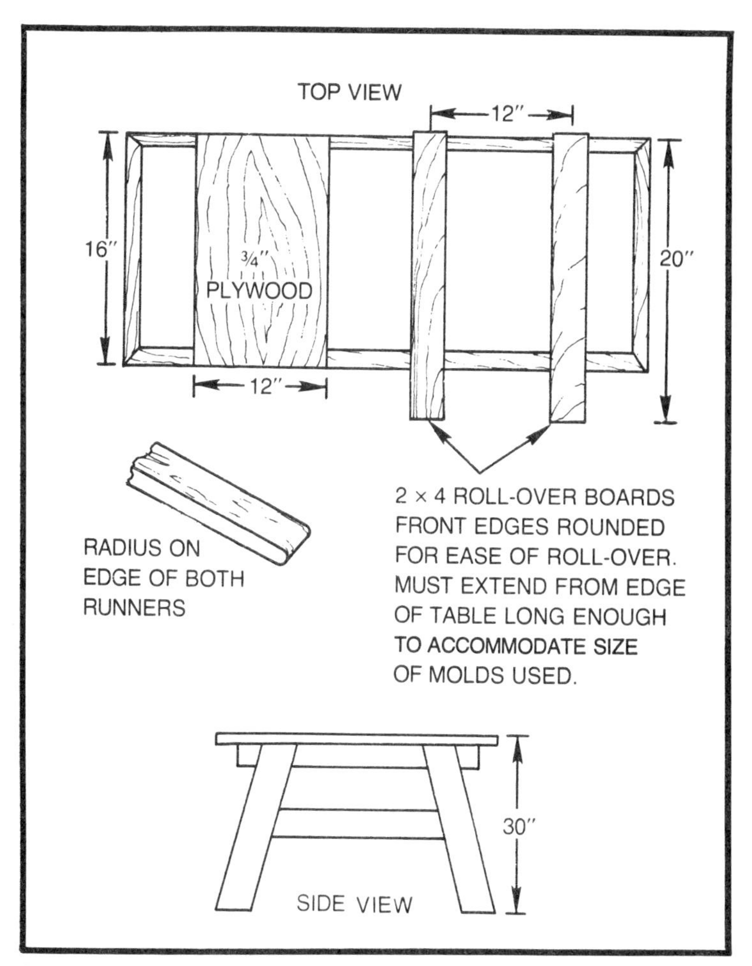

Fig. 7-1. Molder's bench.

The following are the required operations and procedures required in making a sand bench mold and in the correct sequence.

1. Select the correct flask size and depth for the job.
2. Place molding board pattern and drag half of flask on runners of molding bench.
3. Apply dry parting.
4. Riddle sand into flask.
5. Heap floor sand into flask.
6. Peen flask.
7. Ram flask.
8. Strike off flask.

9. Vent flask.
10. Rub in bottom board.
11. Roll drag.
12. Remove molding board.
13. Mark off gating on drag.
14. Apply dry parting.
15. Place cope on drag.
16. Riddle sand into cope.
17. Heap sand into cope.
18. Peen cope.
19. Ram cope.
20. Strike off cope.
21. Vent cope.
22. Cut sprue in cope.
23. Lift cope.
24. Finish cope on set off board.
25. Swab pattern.
26. Rap pattern.
27. Draw pattern.
28. Dry parting in cavity.
29. Cut drag gates.
30. Clean out drag.
31. Close cope on drag.
32. Dust pouring basin with flour.
33. Set mold on floor.

Now you see, there are approximately 33 operations required to produce a simple mold. Exceptions will be noted later for particular cases.

Fig. 7-2. Molder's bench with wheels.

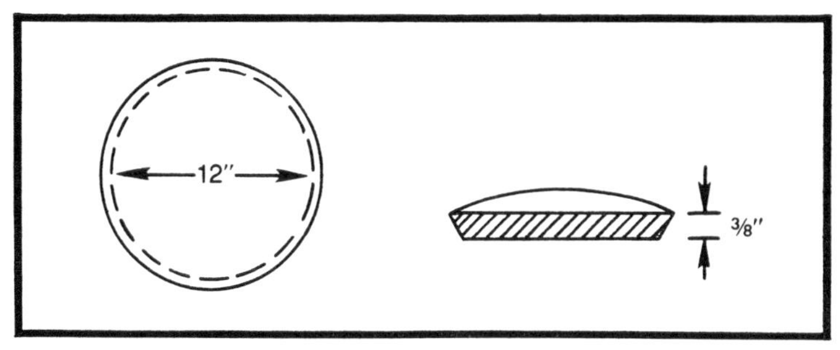

Fig. 7-3. Flat back pattern.

STEP BY STEP

The following exercise will illustrate the various steps as they are followed in the actual casting of a simple flat back pattern. The pattern used is a 12 inch disc, 3/8 of an inch thick. See Fig. 7-3.

Step 1

A flask should be selected that will give sufficient room for gates and enough bearing surface between cope and drag to prevent run outs and sagging which will cause missruns. These missruns leave holes through the casting which are caused by the cope sagging and producing a restricted area. As the metal flows in it will divide and flow around the restricted area and solidify, leaving a hole with rounded edges (missrun). If no missruns show up the casting will be dished on the cope surface. I have seen missruns many times on plaques, because the molder tried to save sand by using too small a flask. In extreme cases the cope will sag and touch off in the cavity. See Fig. 7-4.

This defect can be one or more holes caused by sagging. It is a defect easy to identify. The larger the span the more weight, thus more bearing surface needed.

With a 12 inch flat pattern we should have at least 4 inches from the pattern edge inside of the flask. In our case this gives us a 20 inch square flask.

Now, how much cope depth and drag depth do we need? If the cope is too shallow we will not have sufficient hydrostatic pressure (Sprue Height) to hold the metal against the cope and will wind up with a dished out shrinkage on the cope side of the casting. See Fig. 7-5.

You will learn by experience, but in this case a 4 inch deep cope will give sufficient pressure to hold up the casting.

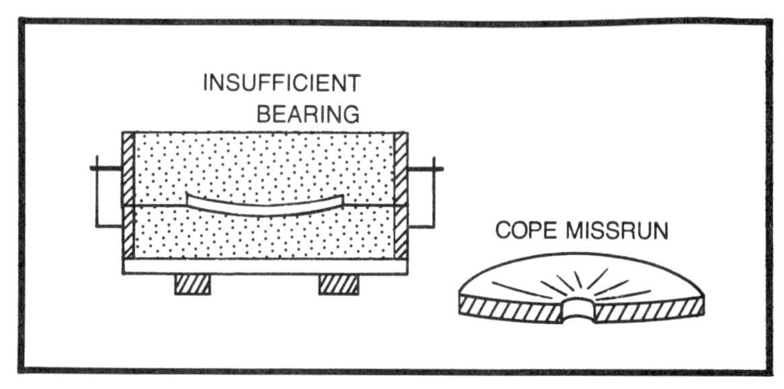

Fig. 7-4. Flask too small.

The two major problems caused by insufficient drag depth are fire and run outs (the metal running out between the cope and drag). The run out is caused by the drag having insufficient depth to hold the weight of the cope and drops away from the cope or does not have inherent strength because the sand body is too shallow compared to its surface area and sags downward. This gives you an opening between the cope and drag at the parting and out runs the metal at the flask joint. Fire: if there is an insufficient depth of sand between the casting cavity in the drag to insulate the bottom board from the heat of the casting during pouring and cooling, this heat can set the wooden bottom board on fire.

That is exactly what happened to Schaff's Gray Iron Foundry in New Orleans. A large flat casting was poured in too shallow of a drag on a big wood bottom board, just at the close of the day. Everybody went home and during the night the bottom board ignited from the castings heat and the whole shop burned down to the ground, everything went but a small out building.

So, we need a 4 inch cope and 4 inch drag 20 × 20 inches. If you take ⅜ inch for pattern depth, we have 3 and ⅝ inches between the cavity and the bottom board.

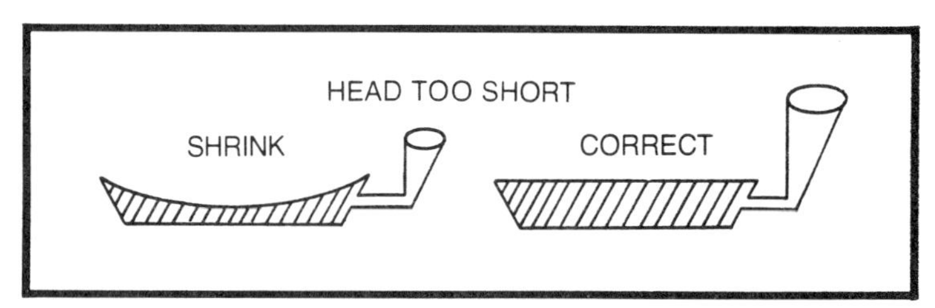

Fig. 7-5. Cope not deep enough.

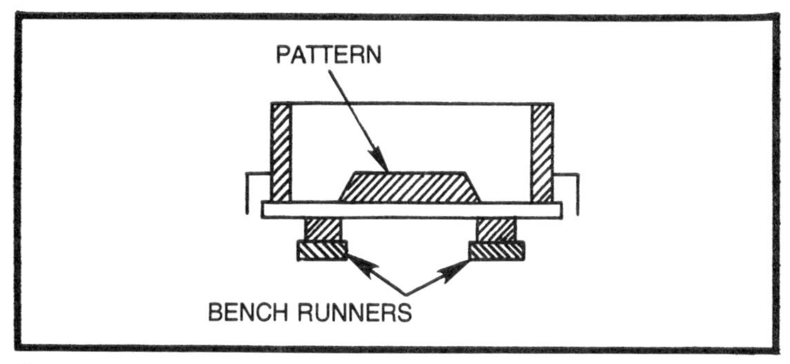

Fig. 7-6. Pattern placed in drag.

Step 2

The molding board is placed on the runners face up, the drag half of the flask placed on the molding board pins down flask parting against the board and the pattern placed in the center of the molding board, cope side parting face down, as determined by the angle of draft (vertical surface taper). See Fig. 7-6.

Step 3

Apply a light coating (very light) of dry parting powder to the inside surface of the flask, board and pattern by giving the parting bag a gentle shake. The bag should be held at least 1 foot above the job and the parting allowed to float gently down on to the job. This gives you a very light coat over a large area. Shaking the bag too close will apply a heavy coat directly under the bag and none or too little elsewhere. Dry parting powder can be purchased under many trade names, Pearl, Superpodium, Non-Sil etc. It is a compound used to prevent the molding sand from adhering to the pattern and board or to prevent rammed sand from adhering to other rammed sand. Sill sand is what we used years ago which is the very dry dust you find on high sills or on top of rafters in the foundry, or in old buildings, barns and the like.

Liquid parting is also available to coat patterns but is not used to part sand bodies. Do not use silica flour as it is very injurious to your lungs and in time will cause silicosis of the lungs (similar to Black Lung). Buy a good grade of non-silica dry parting.

Step 4

Riddle sand into the drag. Place about a half shovel of molding sand in a #4 riddle. The riddle is held parallel to the flask and 6 to 8 inches above. The riddle is held on each side by placing the fingers just under the bottom side of the riddle, holding the riddle loosely, just supporting it on your finger tips. The sand is shook through the riddle by more of a rocking motion of the wrists which causes the riddle to hit the heel of first one hand then the other. This bumping back and forth against the heels of the hands is what does the job, not swinging the riddle back and forth like a pendulum. Handling the riddle wrong can work you to death but with the right motion, it's smooth and easy.

Cover the inside of the flask with about ½ inch of riddled sand. Now with your hands, press the sand down around and on top of the pattern. This helps prevent the pattern from moving out of position.

You riddled the sand over the pattern because this gives you a sand covering the pattern, that is free from clay balls, cat droppings, butts, matches, tramp metal etc., and this clean surface is going to be the face the metal lies against during casting, insurance that you will have a good inner mold face. See Fig. 7-7.

Step 5

Now we fill the drag flask heaping full of molding sand. Now don't do this like filling a wheelbarrow. The shovel full of molding sand is put into the flask by letting it gently slide from the edge in such a manner as to not disturb the pattern position. In other words, don't dump it but place it in where you want it heaping full.

Step 6

Peen the drag. Peening is actually tucking or ramming the molding sand tightly around the inside perimeter of the flask. Here

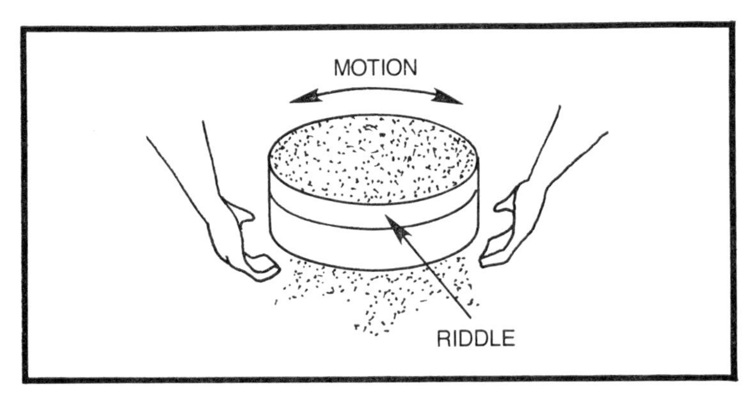

Fig. 7-7. Riddle.

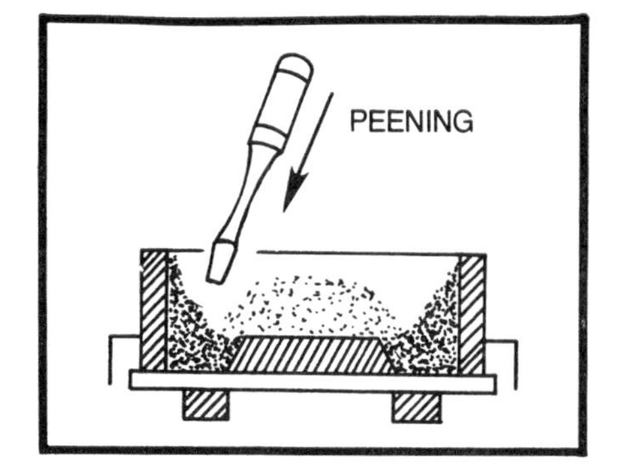

Fig. 7-8. Peening the sand.

you have two choices of tools, the peen end of your bench rammer or the peen portion of your shovel handle. Either is effective. Go all the way around using enough force to pack the sand tightly. Then, shove sand from the middle of the flask into the groove formed by the first peening and peen all the way around again. See Fig. 7-8.

Step 7

Now with the butt end of your bench rammer with the butt face parallel to flask top, ram the sand firmly into the flask adding sand when needed. Ram and add until the mold is firmly packed (rammed) to a point approximately 1 inch above the side of the flask. Keep your rammed sand as level as possible as you work your way up.

Don't try to ram too deep a depth of sand at a time, this results in soft spots (pockets). A 3½ inch depth of loose sand at a time is sufficient to ram then add 3½ inches more. The object is not to ram the mold into a brick but only firm enough and with an even density (mold hardness) throughout the entire mold. The striking force of the rammer should be as if you were cracking walnuts not driving a rail spike into oak. With experience, it will come to you. There are a lot of factors involved, the shape of the pattern the metal being cast etc. A pattern which has pronounced high and low surfaces presents more of a problem in getting an even mold hardness throughout than a flat face, as you have a greater sand depth over the low spots where you must ram harder and less sand depth over the high spots where you must ram lighter. It is something you must learn by doing. See Fig. 7-9.

Step 8

Strike off. The excess rammed sand is now struck off even with the flask with a strike off tool. You don't do this in one shot but strike

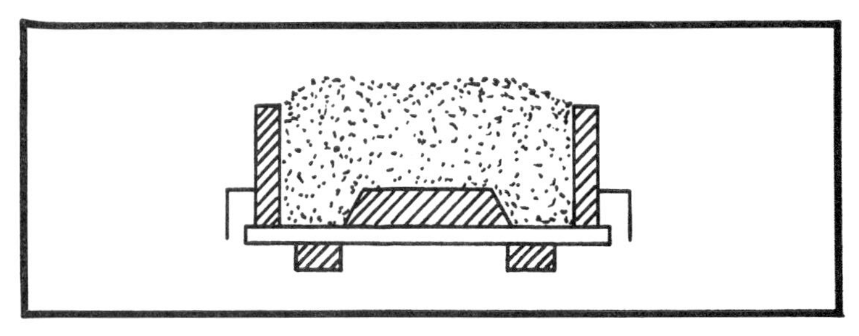

Fig. 7-9. Drag before strike off.

off the sand in increments of about 5 inches at a time. If you start at one end and try to take it all off in one scraping you will not only wear yourself out, but might take out some large chunks below the flask sides. See Fig. 7-10.

Step 9

Vent drag. With a ⅛ inch diameter vent wire, vent the drag generously, try not to hit the pattern but if you do no harm is done. See Fig. 7-11.

Step 10

Rub in bottom board. Due to the fact that most bottom boards are uneven and they become burned and warped by use and rough handling, they will not support a mold evenly. If the joint between the bottom board and the mold is not a good complete and even joint, the sand in the drag will crack, distort and push up when the mold is closed and weighted. A cover layer of loose molding sand, a minimum of ½ inch thick is spread over the struck off drag and the bottom board is rubbed into this bed until a firm seat is formed. Most

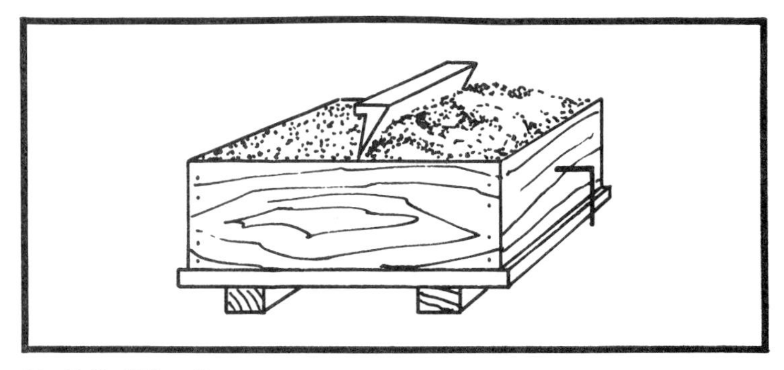

Fig. 7-10. Strike off.

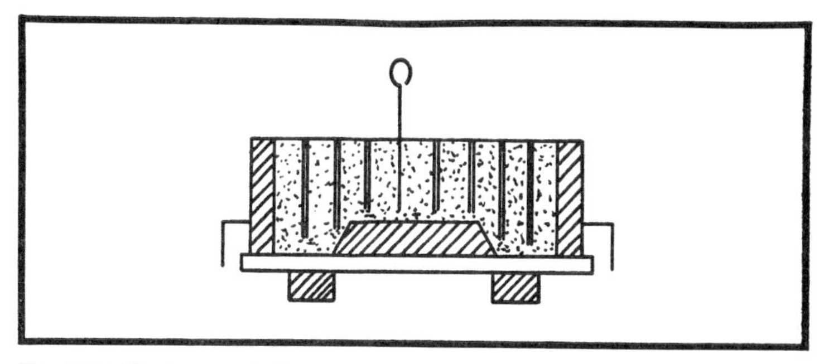

Fig. 7-11. Placing vent holes.

molder's when they have finished, lift the board off to check and fill in any low sports, if any. See Fig. 7-12.

Step 11

Now the drag bottom board and molding board as a unit must be rolled over. If done correctly, it is a smooth quick action. If not, it can

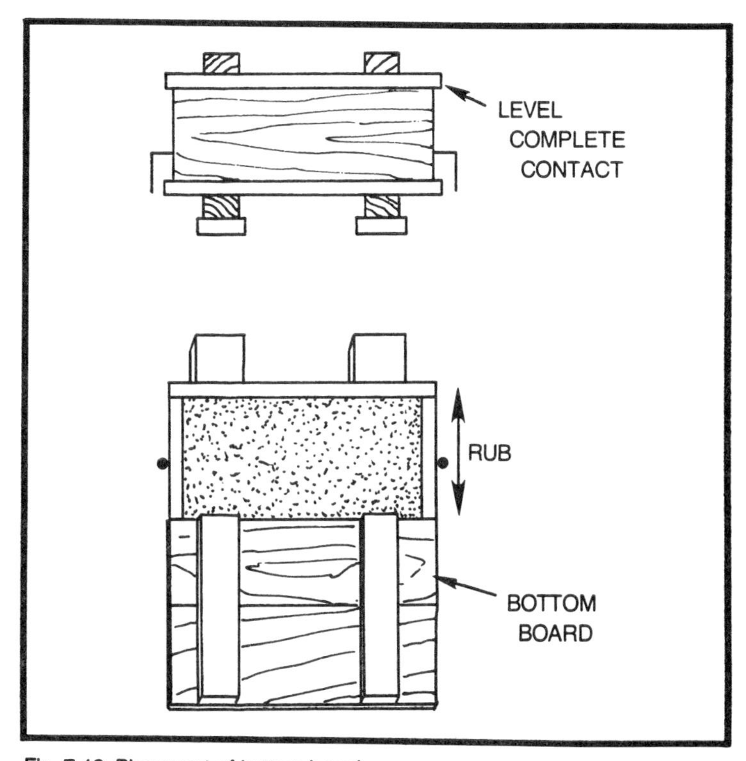

Fig. 7-12. Placement of bottom board.

break your back, or you can lose it and find the whole works on your feet. Once started, the action must be continuous (one motion). If you stop anywhere along the line you will lose it.

With the arms parallel to the ends of the flask, grab the flask and molding board between the hands. Remove the flask from the bench and place the cleats of the bottom board under the 2 × 4 inch runners on the molding bench, using the rounded ends of the bench runners as a fulcrum and roll the flask over. As stated before, once you start don't stop. One complete uninterrupted motion. Practice with a flask filled with sand. See Fig. 7-13.

If the flask is too large or heavy to handle make a two man roll. If not too heavy but drag too deep for comfort, clamp with a C-bar clamp and wedge before rolling over. The pressure against the underside and ends of the bench runners is all that keeps the bottom

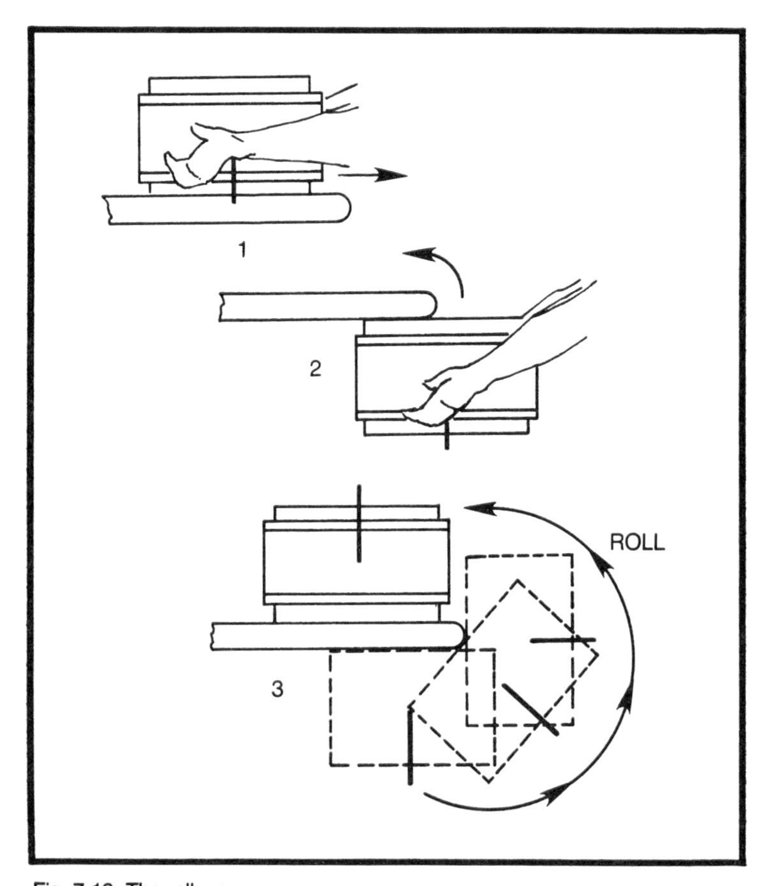

Fig. 7-13. The rollover.

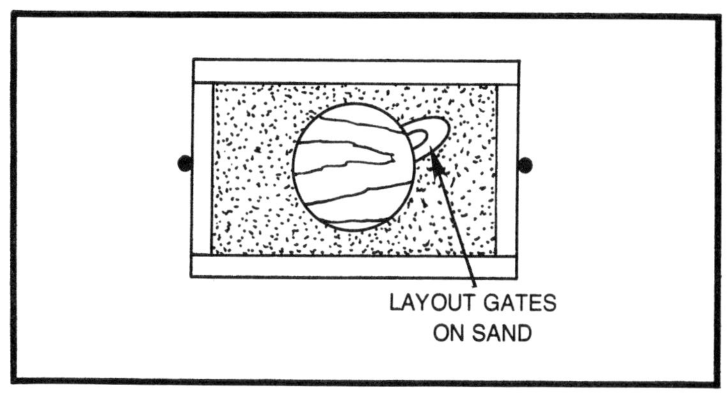

Fig 7-14. Placing gates.

board in place during the roll over. Slack up or stop and you will catch the bottom board across the top of your feet.

Step 12

Remove molding board.

Step 13

You now have the drag mold parting face upward. Now in the sand, lay out the gates with the blade end of yankee filter. Simply draw these into the sand with the narrow edge. The reason for this is that when the cope is rammed this lay out will show on the cope and locate where to cut the cope runner over to the sprue. See Fig. 7-14.

Step 14

Apply dry parting to drag face.

Step 15

Place cope on drag. Now repeat steps 4, 5, 6, 7, 8, and 9 to the cope half thereby completing steps 16, 17, 18, 19, 20 and 21 respectively.

Steps 10, 11, 12, 13, and 14 apply to drag flask only.

Step 22

Cutting sprue. The sprue is the opening down channel in the cope sand which directs the poured metal into the runner thence into the gates which lead into the mold cavity. See Fig. 7-15.

We have two choices with the sprue, we can use a set sprue (sprue stick) which is set in the cope when ramming the cope and

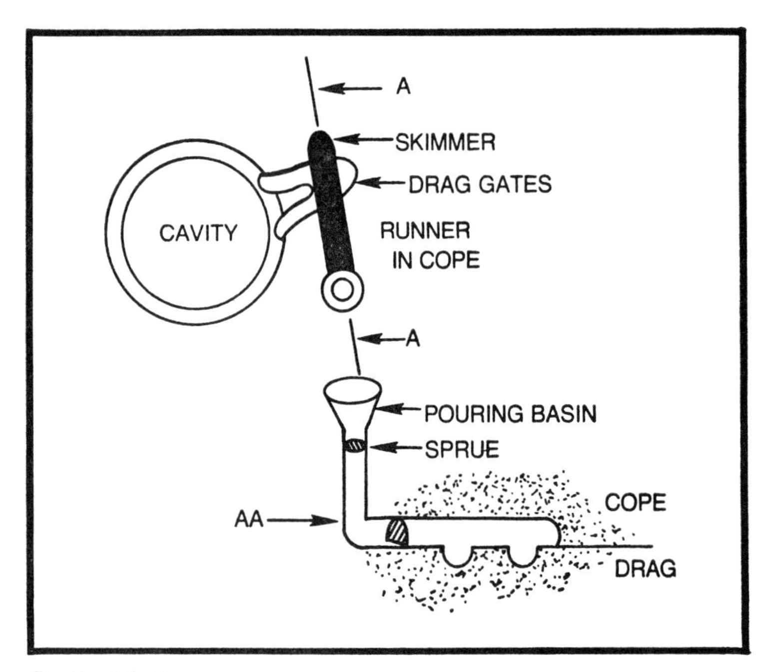

Fig. 7-15. Cutting sprue.

drawn out to leave the sprue hole or we can cut the sprue with a tapered tubular sprue cutter, which is the better way.

We now have the cope completely rammed and vented. Now we noted, I hope, where our gate lay out was on the drag before ramming the cope. The sprue cutter is held against the side of the cope (the cutting end at the flask parting) your thumb even with the top of the cope. This gives us our cut depth. Now without moving your thumb the sprue is cut (punched) down through the sand in the cope until we hit our thumb. See Fig. 7-16.

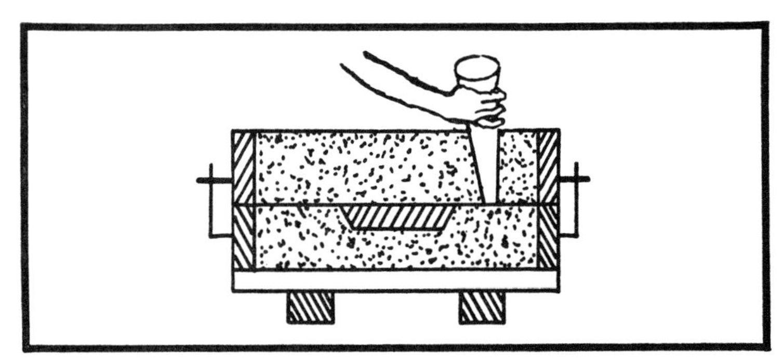
Fig. 7-16. Use of sprue cutter.

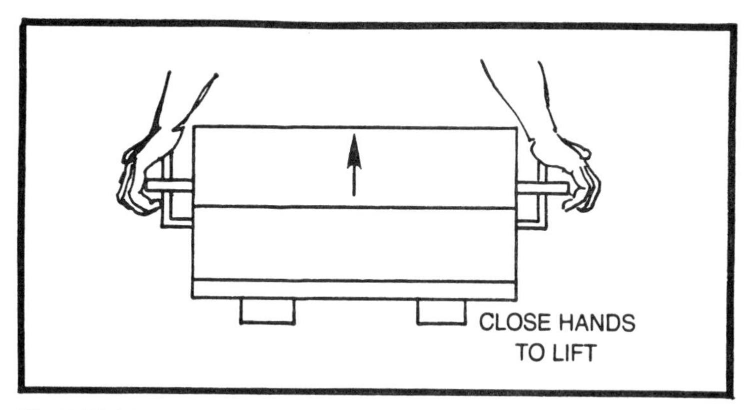

Fig. 7-17. Lifting the cope.

Step 23

Lift cope. The cope is lifted off straight. This is accomplished by placing the fingers under the cope guides and the heel of your hand or your thumb on top of the drag pin, whichever best. As this distance varies with flasks you must choose thumb or heel. If the drag pin only protrudes a short way through the cope guide, you have to use your thumb. The ideal length of pin through the cope guide is, when the heel of your hand is on the top of the drag flask pin, and your fingers under the cope guide, your hand is only cupped slightly.

The cope is lifted straight up by closing both hands simultaneously. A straight lift is not needed for a flat back pattern, however, if you are picking up a pocket of sand on the cope or coming off of a pattern that projects into the cope, you must come straight up and off, always. See Fig. 7-17.

Step 24

Finish the cope. The cope is set on its side on the set off portion of your molder's bench, the 12 × 16 inch plywood. At the parting face of the cope we see the sprue hole and the transfer of our drag gate lay out. With a gate cutter we cut a ½ round channel in the sand of the cope from the sprue across the transfer lines. This is our runner. We slick the sand down smooth in the channel and fillet the junction at the sprue and end of our runner with our finger. On the top of the cope we cut the sprue opening into a funnel shape to form a pouring basin. Now we blow away with molder's bellows any loose sand from both faces of the cope and through the pouring basin and sprue. The cope is now finished.

The reason we completely finish the cope first without touching the drag is simple. If we have a bad cope or damage it beyond saving, we can shake the sand out, put the cope back on the drag, and ram a new one. If we had drawn the pattern and finished up the drag then ruined the cope, we would have to start the complete mold over. See Fig. 7-18.

Step 25

With a bulb sponge we slightly dampen the sand around the pattern perimeter to prevent the edge from breaking away when we draw the pattern.

Steps 26 and 27

Rap and Draw Pattern. Screw a draw pike into the center of the pattern, and rap the pattern in all directions with a rapping tool, and carefully lift the pattern straight up and out of the drag. Repair any damage to the cavity with your double ender. See Fig. 7-19.

Steps 28 and 29

Dust and cut drag gates. Now before we do anything else, we shake dry parting powder in the mold cavity. We do this because when we are cutting the gates, the sand might, and some will, fall into the mold cavity but will not stick because of the parting, and when we blow the drag out, it will come completely clean with a minimum of blowing.

We now cut a ½ round channel, the gates, where we made our gate layout. You never cut into a cavity but cut away from it, so

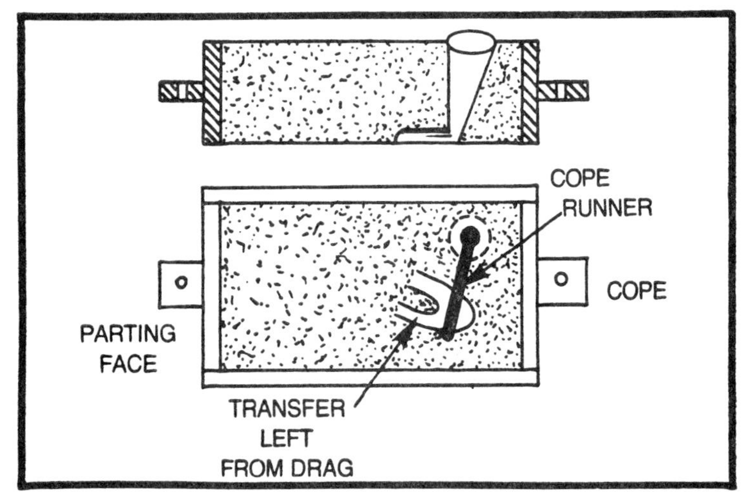

Fig. 7-18. Finished cope.

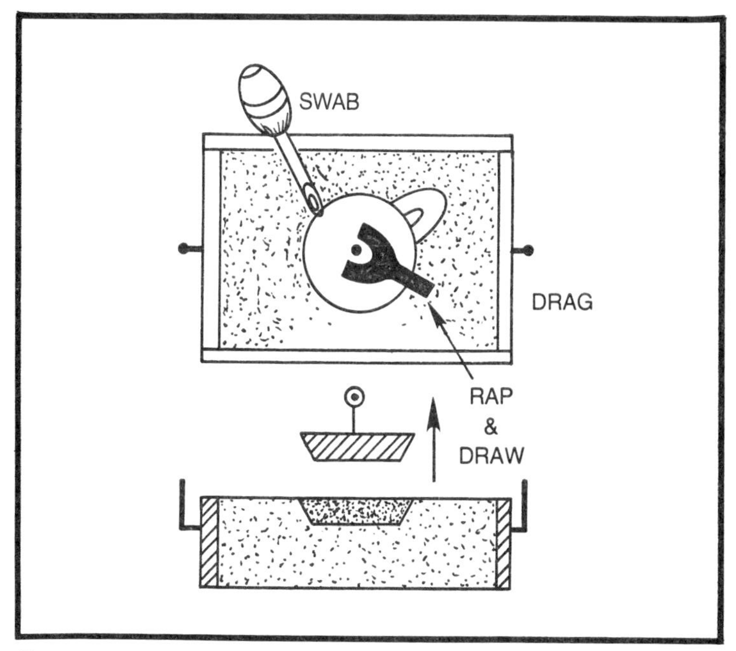

Fig. 7-19. Rap and draw of pattern.

as not to break off any sand either side of the gate making the gate entrance larger than we wish. Fillet the gates where they enter the cavity with your finger as we did with the runner in the cope. Finish the gates smooth with no sharp edges or loose spots. See Fig. 7-20.

Steps 30, 31, and 32

Clean drag, close cope on drag, dust pouring basin. The drag is blown out with the bellows good and clean. Now we are going to close up the mold called coping or closing. It is here where more often than not, loose sand falls into the mold cavity from off of the cope flask sides, guides etc. To prevent this, we first check the cope, pouring basin, sprue and runner. Blow off both sides again with the bellows. Now, never move a cope over a finished drag any other way than parting face to parting face and parallel.

Get the cope in this position before you are over the drag, then move in and close the mold. If this rule is followed, any sand on the flask sides and ends will fall off before you are over the drag. Now close with the same hand and finger action as lifting, but in reverse. The hands are closed, the guide hole covered with the heel of the hand.

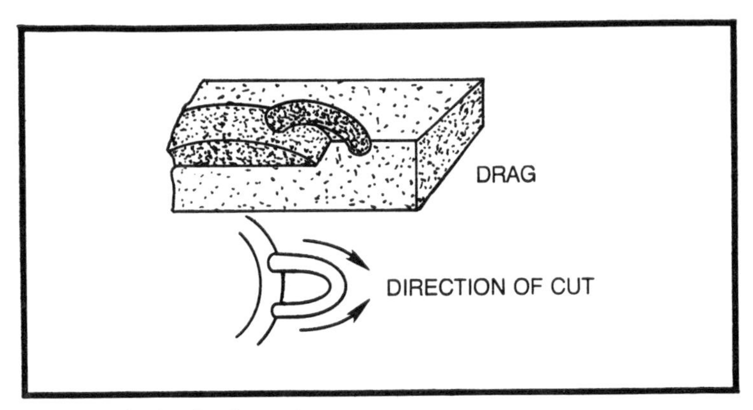

Fig. 7-20. Cutting the drag gates.

When you feel the drag pin come up through the guide and touch the heels of both hands, level out top of cope and lower cope on to drag by opening both hands evenly and slowly until the cope seats on the drag. Don't drop the cope on to the drag or bump it down when closing. With a parting bag, or sock, full of wheat flour shake enough into the pouring basin to give it a white color. This does two things: (1) the dampness from the sand will stick to the flour and help prevent sand washing into the mold from the pouring basin when the mold is poured. Some molders dust all gates and runners with flour; (2) the pouring basin being white makes a much more visible target to hit when pouring the mold (a must when pouring late in the evening or at night). An unfloured pouring basin can simply disappear in a shadow if the lighting is bad or appears to be where it is not. See Fig. 7-21.

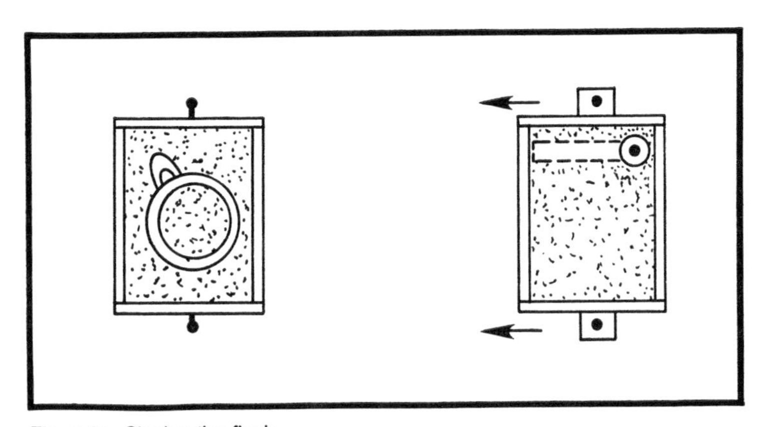

Fig. 7-21. Closing the flask.

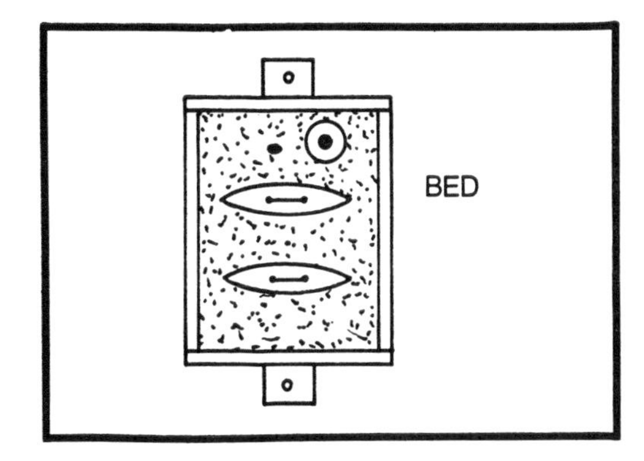

Fig. 7-22. Sand bed.

Step 33

Because of the uneveness of a foundry floor and the uneveness of the cleats on the bottom board, a bed of loose floor sand is fanned out on the floor with the molding shovel and leveled off with the leading edge of the shovel. The mold is then set off on this bed, not rubbed into the bed. The mold is set down easily, not bumped. See Fig. 7-22.

BENCH PRACTICE

In all bench or floor molding the 33 steps we just went through are more or less carried out regardless of what we are molding. We will now deal with some bench practices involving different types of patterns. We will only detail that part which would differ or be an added step to the 33 steps already covered.

Bevel Gear Blank Pattern

In Fig. 7-23 we see a cross section of a small bevel tooth gear blank which will have to be parted on the plane shown by the dotted

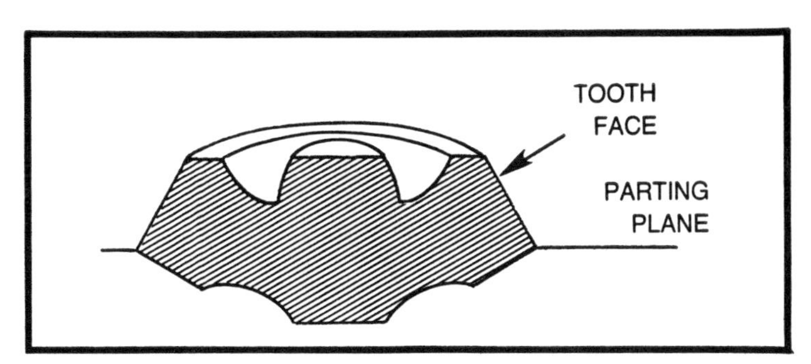

Fig. 7-23. Bevel gear blank pattern.

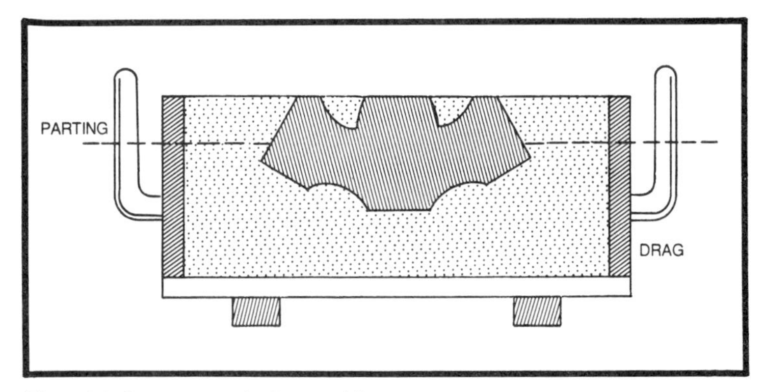

Fig. 7-24. Gear pattern in drag mold.

line. The pattern is placed on the molding board in the drag flask with the face up that is to get the machined teeth, so when rolled it will be cast in the drag. The reason that we want this face to be cast in the drag is the face of a casting that is produced (cast) in the drag half of a mold stands a much better chance of being clean and free from dirt and other defects than the cope face. If some dirt, slag or sand comes into the mold during pouring it will float on the metal and wind up 90 percent of the time on top of the casting, and as we wish to machine teeth on face "A", we will cast it in the drag. The drag is rammed in the usual manner and rolled over. See Fig. 7-24.

We have face "A" in the drag but we also have the parting plane rammed in the drag below the parting face of the flask. We must cut the sand away, all around the parting line. This is called coping out. In doing this the portion of sand coped out becomes part of the cope. See Fig. 7-25.

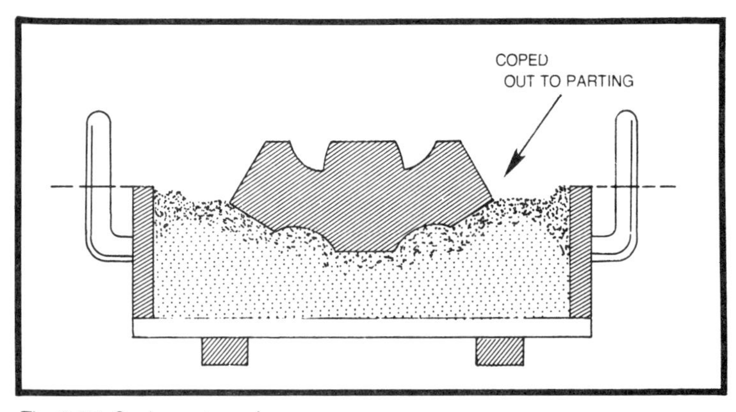

Fig. 7-25. Coping out sand.

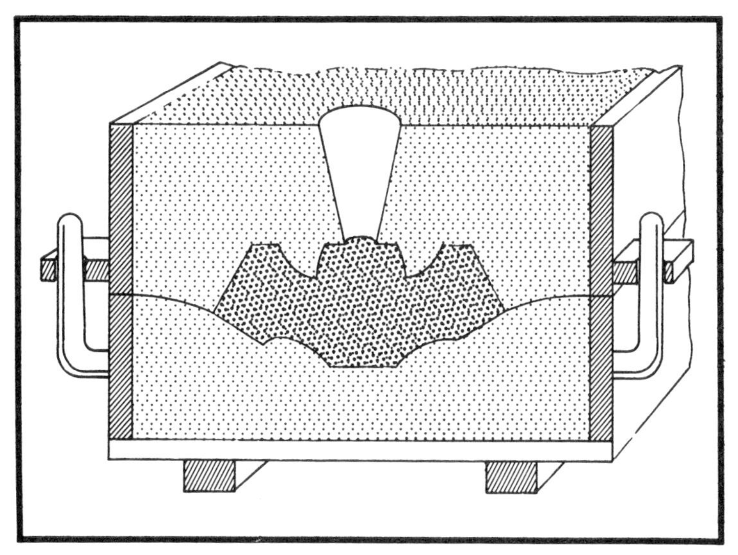

Fig. 7-26. Finished mold.

When coping out, the sand is cut back on a flat plane from the parting line on the pattern, and then sloped up to the flask parting so that the body of the sand which is projecting from the cope is not sharp (which could easily break away).

The coping is done with the spoon end of a spoon and slick tool. The sand is slicked down to form a smooth parting surface. The drag is blown out with the bellows and the cope flask put on.

We dust the drag with parting powder and place a sprue on the hub of the pattern and finish making the cope in the usual manner. Lift cope and finish drag. Draw pattern and finish cope, and close mold. See Fig. 7-26.

Shoring Washer Green Sand Core

The shoring washer is molded like any simple flat back pattern with the exception that before the cope is rammed the body of green sand forms the core through the washer is pressed down to tighten it up somewhat, concaved on top and a headed nail pushed down through its center. This assures that not only is the core firm (rammed tight enough) but that it will not come loose when rapping and drawing the pattern, and stay down when blowing out the drag with the bellows. Small green sand cores can if not nailed be blown loose easily. The cope will hold it in place during pouring. See Fig. 7-27.

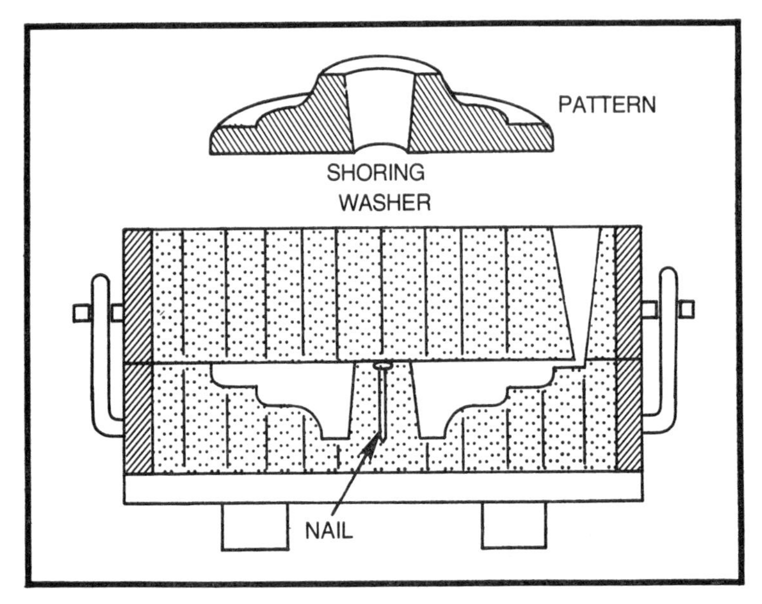

Fig. 7-27. Shoring washer mold.

Flat Faced Pulley, Split Pattern

In Fig. 7-28A we show a cross section of a split pattern for a flat faced pulley with a cored hole through the hub.

The cope and drag halves of the pattern are pinned together with pattern dowels, female in drag half and male pins on cope half. The core print on the cope half of the pattern you will note is tapered and the drag print is not. The cope core print is tapered, matching a taper on the core. The cope print leaves a tapered cavity in the cope mold, which because of its taper will line the core up parallel when

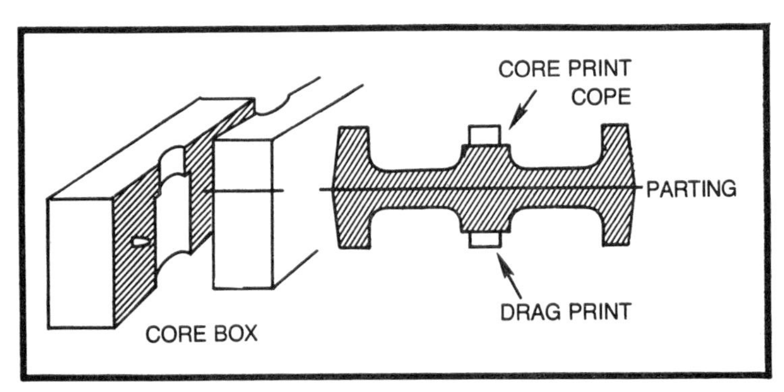

Fig. 7-28A. Development of split pattern mold.

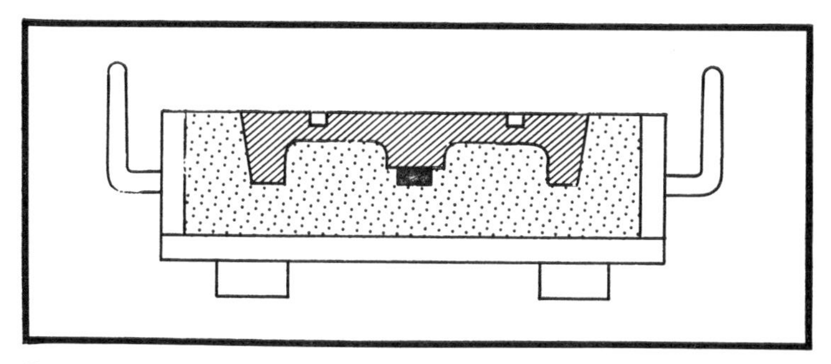

Fig. 7-28B. Drag mold.

closing the mold without shaving sand into the mold which would happen if the cope print were straight sided like the drag print. Both cope and drag prints are usually tapered on large cores.

In molding the pulley the drag half is placed on a molding board in the drag flask, and rammed, vented and rolled over on a bottom board in the usual manner. The cope half of the pattern is placed on the drag half. A sprue stick is placed on the hub, the cope rammed, making sure the pocket between the inner face of the rim and hub is rammed properly. The sprue stick is removed and a molding board placed on top of the cope. The cope is lifted and rolled over on its back on the molding board on the set off bench. We now have the cope in a position where we can swab and draw the cope half of the pattern which stayed with the cope when we lifted and rolled it over.

It is much easier to draw a pattern half from the cope, without damage, with the cope on its back. Drawing the half pattern from the cope with the cope on its side is impossible most of the time, because drawing the pattern out on a horizontal plane is tough to judge. Lifting the cope from the cope half of the pattern can be tougher yet, should the cope half of the pattern stick down to the drag half and not come with the cope. Plus the fact that if the cope half of the pattern comes with the cope (stays in) we don't have to worry about a straight cope lift. We are lifting a flat plane from a flat plane. The drag half of the pattern is swabbed etc., and drawn and the drag finished. The cope half of the pattern is swabbed, etc., and drawn from the cope. The cope is now set on its side to finish the sprue and pouring basin. Don't forget to shake dry parting into the cope and drag cavities after the patterns are drawn. Before doing anything else see step 28 of the 33 steps in molding and study Fig. 7-28, A through D.

The core is set in the drag. The mold is now closed and set on the floor for weighting and pouring.

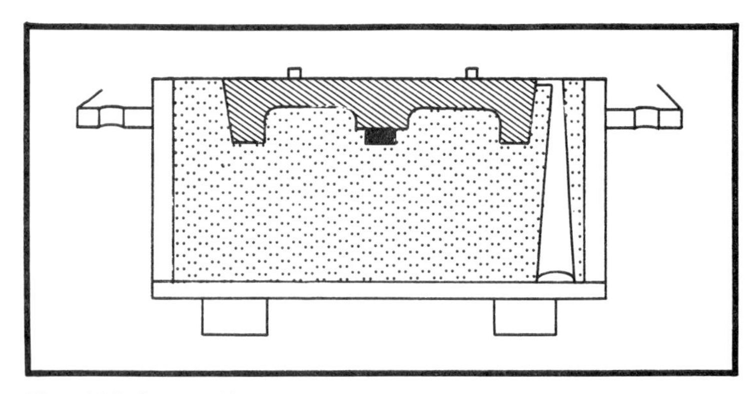

Fig. 7-28C. Cope mold.

Now, should we have a split pattern that because the weight of the cope half of the pattern and its shape will not come up with the cope, and if so, it could fall out of the cope before we get the cope over on its back. The chances of a heavy half pattern in a cope falling straight out of the cope without damaging or wrecking the cope is very slim. If we have reason to suspect that the pattern half in the cope will not come up with the cope (stick down, fall out etc.), we must assure that it does by hooking it to the cope. To do this we first ram up the drag half, roll it over, put the cope half of the pattern on the drag half, place the cope flask on the drag flask. Now we screw one or more come along pins into the top of the pattern.

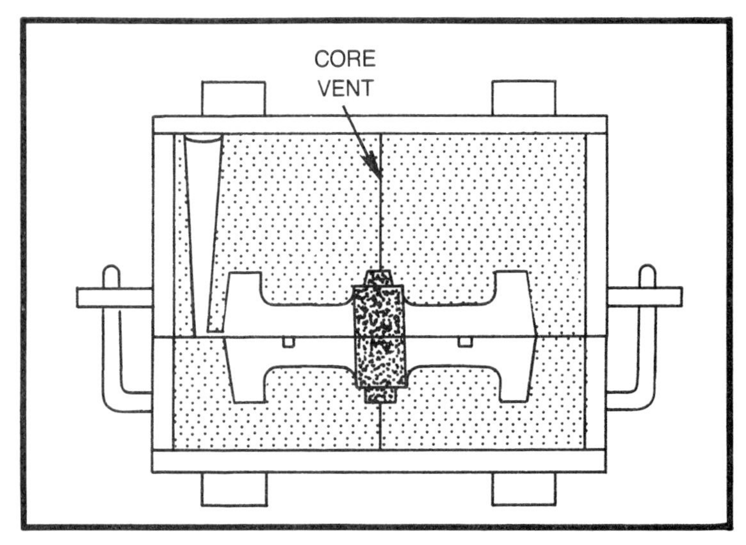

Fig. 7-28D. Complete mold.

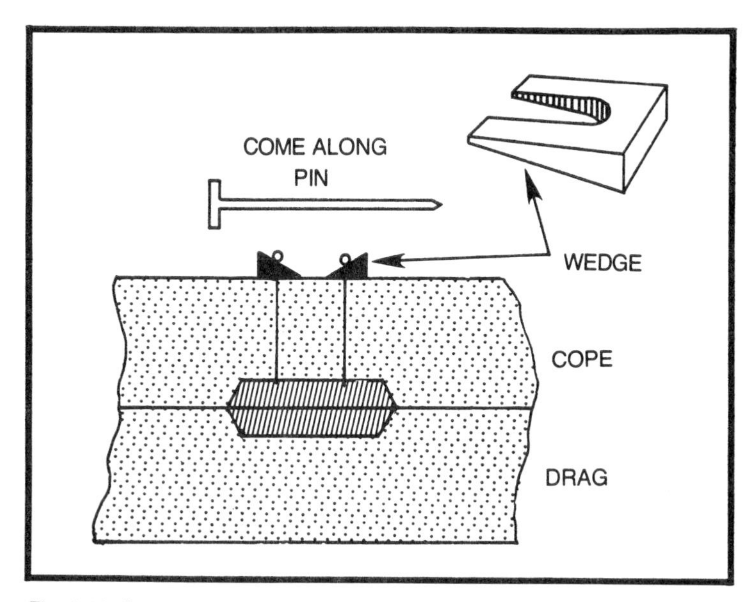

Fig. 7-29. Come along pins.

The come along pins are sticking up above the cope, the cope is rammed in the usual manner. Place a wedge between the underside of the "T" handle of the come along pins and the sand top of cope. The pins, each secured by a slot wedge, will carry the pattern up with the cope. The cope is set on its side and the wedges removed, the come along pins unscrewed out and removed, a molding board placed against the cope and the cope eased over on to its back. Light patterns are often carried up with the cope by sharpened welding rods driven into the pattern and held against the cope (top) by grabbing the rod with a pinch type clothespin at the junction of sand and pin. See Fig. 7-29.

This method is used on both bench and floor work. In very large work under the crane, the pattern is often wired to the cope bars and when rolled over on horses the wires are cut loose from below. See Fig. 7-30.

Flat Faced Pulley on Match Plate

Here we have the same pulley we molded with the split pattern only mounted on a match plate.

The match plate is placed between the cope and drag flask, held in position by its guides through which the drag pins pass, cope pattern is in the cope flask and the drag pattern is in the drag flask.

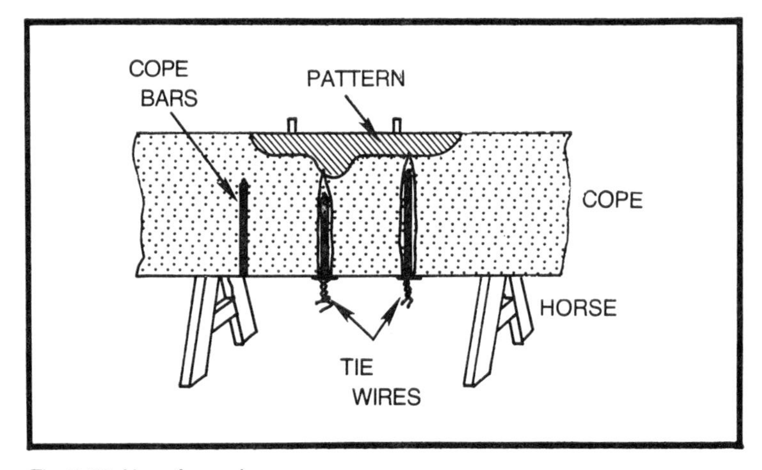

Fig. 7-30. Use of cope bars.

The flask is placed drag up on the molding bench. The drag is rammed in the usual manner, a bottom board rubbed in and the whole assembly the rammed drag flask, the match plate and empty cope rolled over. The cope is rammed, vented and the sprue cut. Now we have the two flask halves rammed up withe match plate between them. Now we stated before it is difficult to lift a cope off a pattern it is best to lift the pattern from cope. Here we must lift the cope from

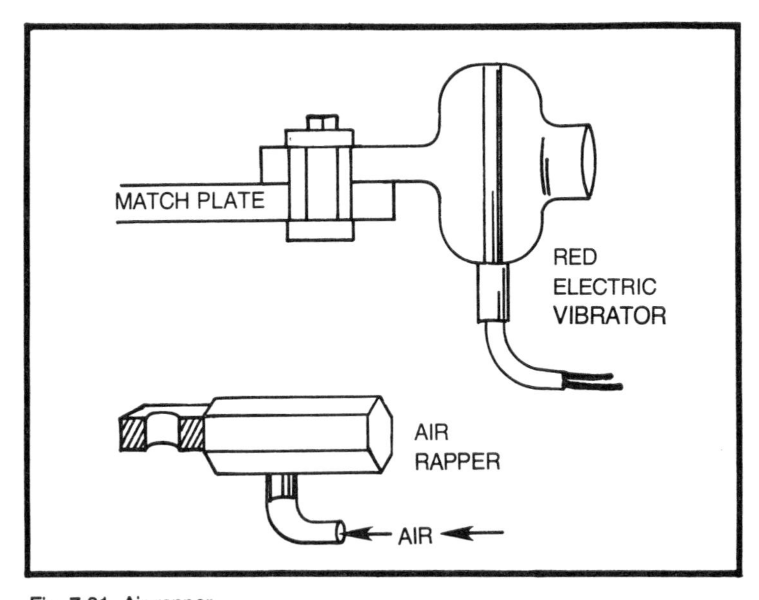

Fig. 7-31. Air rapper.

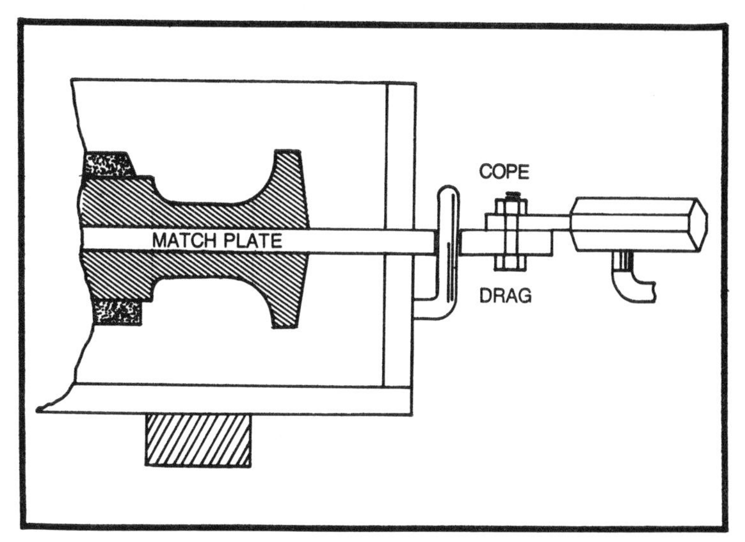

Fig. 7-32. Using air rapper.

the pattern, no choice it is hooked down. Also, we cannot swab the pattern. The big advantage we have is the pins of the flask help us guide the cope straight off and the match plate straight off of the drag. Before we lift the cope we can rap the edge of the match plate with a mallet or bar, this will help some but not much. What we need is called a match plate rapper or vibrator. There are two types: air and electric. The rapper is held to the match plate by a bolt which passes through the match plate and the clamping ear on the rapper. See Fig. 7-31.

The rapper is operated by a knee switch or a knee operated air valve for the air rapper.

To draw the cope start the rapper and lift the cope. The rapper is then stopped. It is only kept running until we clear the pattern. This continuous vibration during the lift shakes the pattern loose from the sand, letting the cope lift clean. The cope is set off on its side and finished. The match plate is then rapped and lifted off and set aside, the drag blown out, core set and the cope placed on the drag. See Fig. 7-32.

Pattern Requiring an Irregular Draw

In this case we have a pattern which must be drawn on an angle or arc (not vertically) in order to remove it from the sand without damage to the mold. See Fig. 7-33.

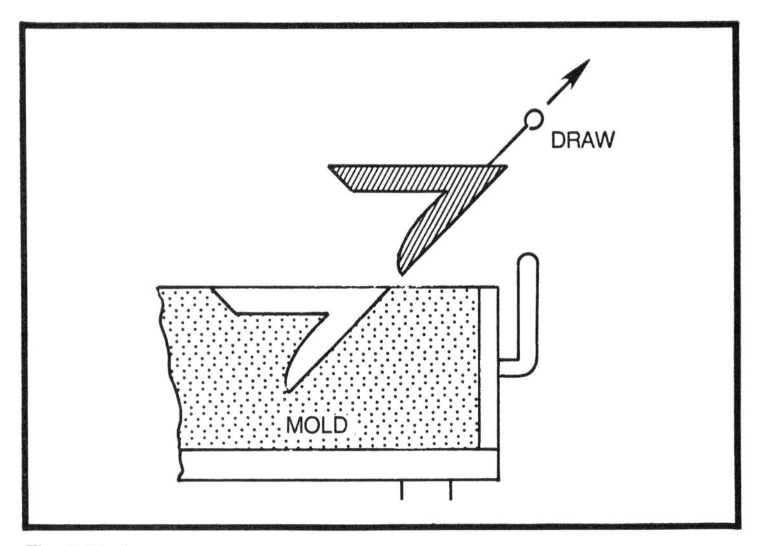

Fig. 7-33. An angle draw.

Molding a Flanged Drum Casting

Here we have a three part pattern consisting of a bottom flange, body and a top flange. See Fig. 7-34.

The bottom flange with core print is placed on a molding board in the drag. (Drill a hole in the board to clear the pattern pin on bottom flange.)

Ram the drag up and roll it over on a bottom board. Dust the drag with dry parting. Place the body pattern on the flange pin, and

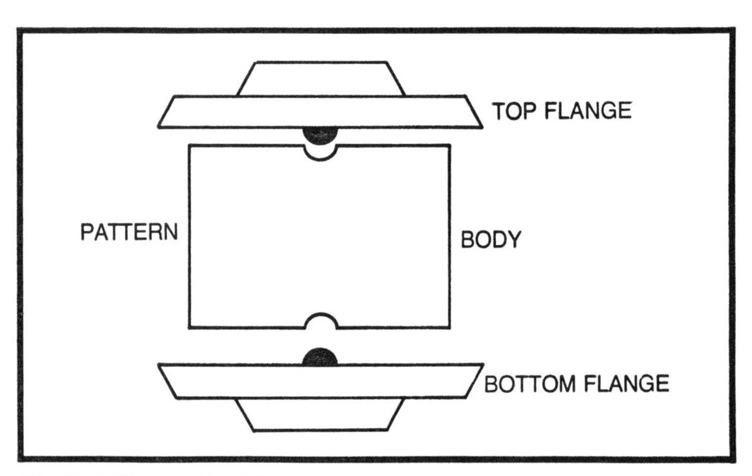

Fig. 7-34. Three part mold.

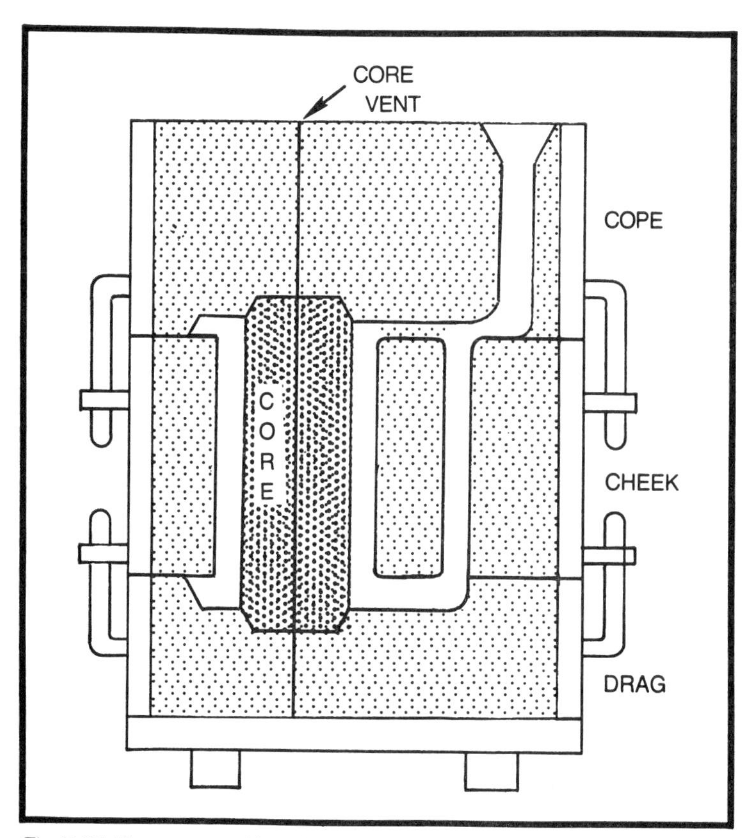

Fig. 7-35. Three part mold ready for pouring.

place the cheek flask on the drag. Ram up the cheek, striking it off even with the top of the body pattern. Dust with dry parting, place top flange on body pattern and cope on cheek. Ram cope place vents and gates. Lift off cope, draw cope pattern and finish cope. Draw pattern from cheek, then lift cheek from drag. Draw drag pattern from drag. Place core in drag print, put cheek on drag (careful here, don't hit cheek sand). Place cope on cheek weight and pour. See Fig. 7-35.

Grooved Pulley Molding in 2 Part Flask

To mold this type of pattern (see Fig. 7-36) in a two part flask, without a cheek, we have to make what is known as a false cheek to take care of the groove in the pulley. Place one half of the pattern and the cope on a molding board with a sprue pin on the hub. See Fig. 7-37.

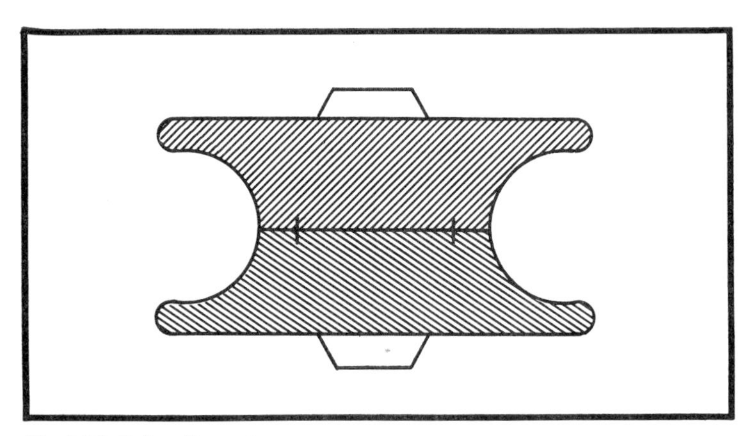

Fig. 7-36. Split pulley pattern.

Ram up the cope vent and remove sprue and roll cope over on a molding board. Then, make a parting (cope out) down to the bottom of the half groove and place drag half of pattern on cope half of pattern. See Fig. 7-38.

Sprinkle parting powder on parting you cut. Now tuck and ram sand into the groove in the pattern and the coped out parting, making a parting at the top of flange down to the flask parting. See Fig. 7-39. We now have a green sand ring or green sand groove core.

Dry parting is shaken over the flask parting and the green sand ring core, then the drag flask is put on and rammed up. The drag is lifted off, the pattern section D is removed. See Fig. 7-39, the drag is replaced, which will support ring core. The entire mold is now rolled over cope up. The cope is drawn and the other pattern section is removed, part C see Fig. 7-39. The cope is finished, the drag blown out, then set the hub core and close. See Fig. 7-40.

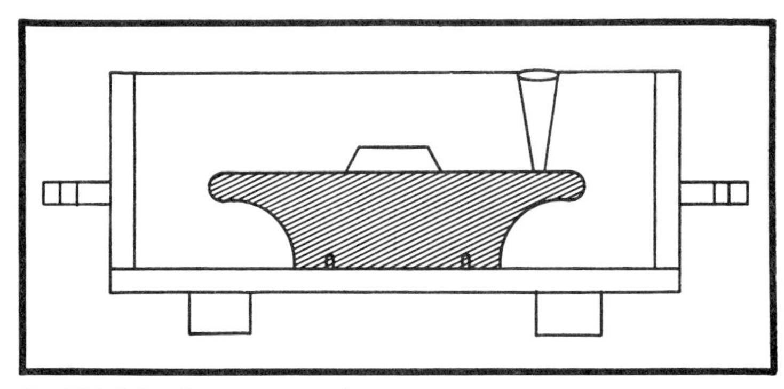

Fig. 7-37. Split pulley pattern, step 1.

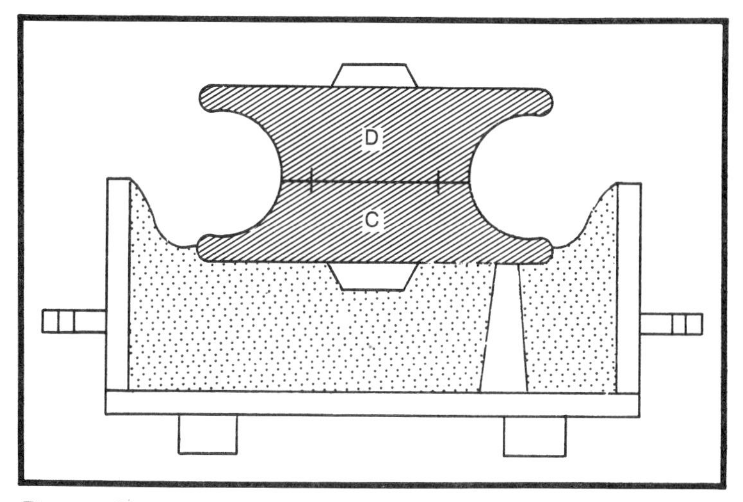

Fig. 7-38. Split pulley pattern, step 2.

The same casting can be made with a dry sand ring core and with a core print on a solid pattern, molded all in the drag. See Fig. 7-41.

Double Rolled Pattern

In this case we have a pattern with a deep pocket (a bowl shaped pattern). If molded in the usual way, the cope half of the finished mold would have the sand that cores out the interior of the casting hanging down from its face. This could break or fall from the cope

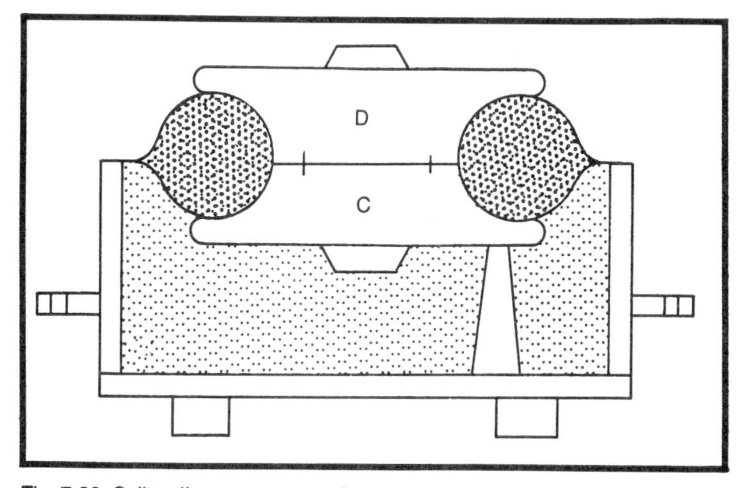

Fig. 7-39. Split pulley pattern, step 3.

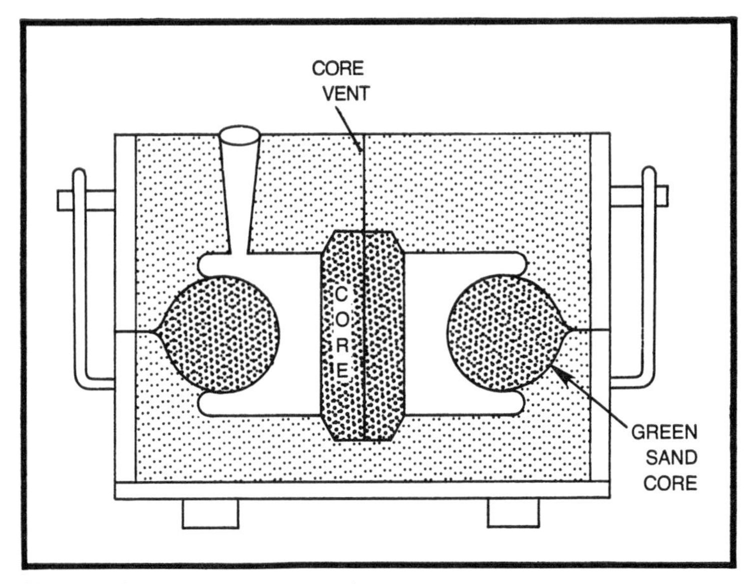

Fig. 7-40. Split pulley pattern, step 4.

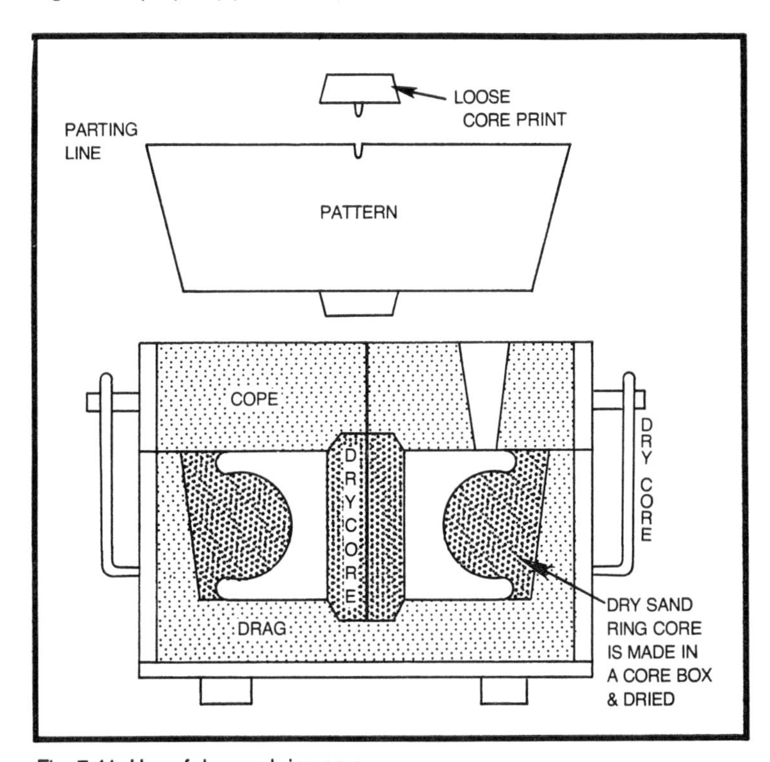

Fig. 7-41. Use of dry sand ring core.

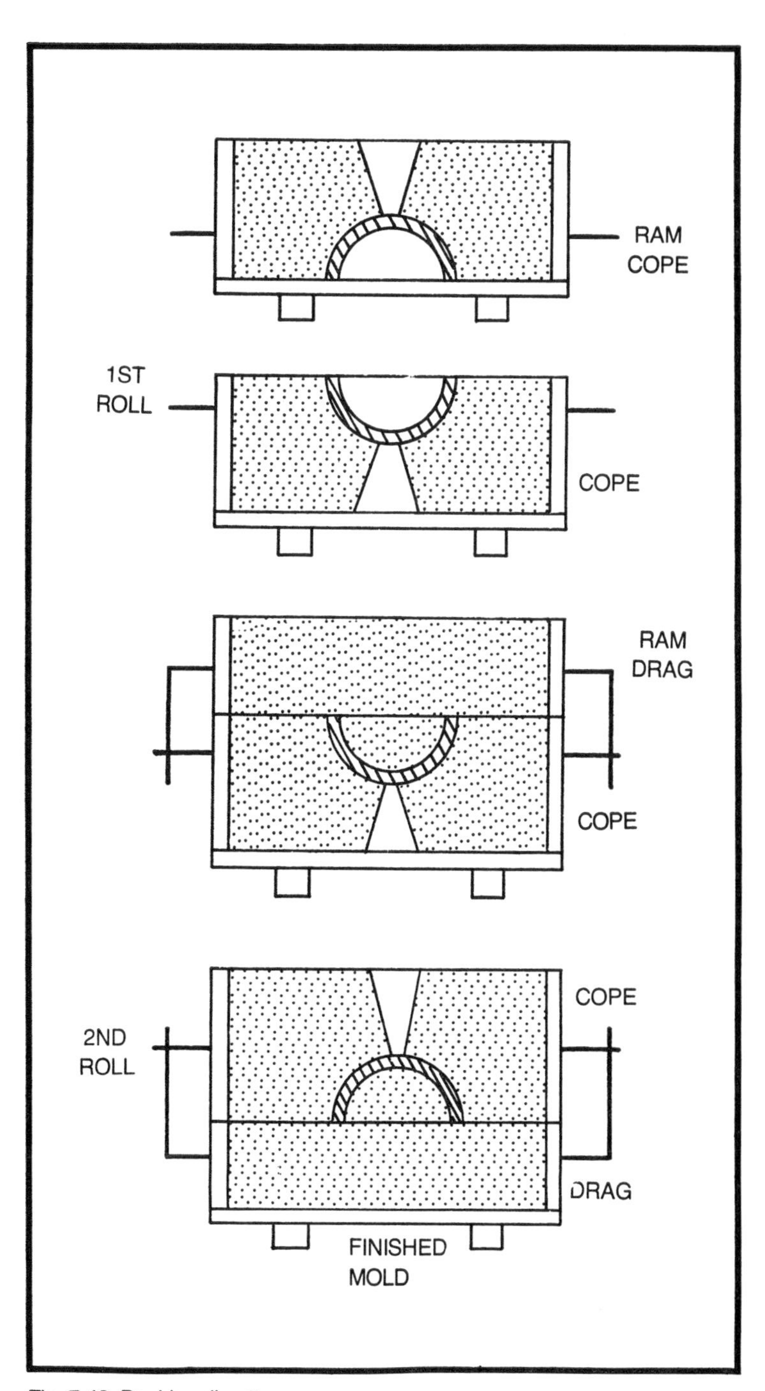

Fig. 7-42. Double roll pattern.

(due to its weight) into the drag during molding, weighting or pouring the mold. The safest way to make the mold would be to wind up with this deep core resting on the drag upright in place of hanging from the cope.

In order to do this we must ram the cope first, roll it over on its back. Ram the drag, roll over cope and drag together, putting the core standing on the drag (Double Roll). See Fig. 7-42.

Pattern Used as Dry Sand Core Box

Now let's go back to the bowl shaped pattern.

Making the Core

The interior of the pattern (cavity) is dusted with core box parting and rammed full of core mix then struck off even with the parting face. At this point we imbed in the core two or more lifting hooks, made of wire. The hooks are imbedded in the green core, their tops even with the parting face, then you scoop the core sand out to clear the openings of these hooks. The core is vented, a core plate placed on the parting, box and plate rolled over. The box (in this case also the pattern) is rapped and drawn off. The core is then baked (dried). See Fig. 7-43.

To make the mold the drag is rammed and rolled over. The dry sand core is placed in the cavity of the pattern. A U-shaped wire is passed through the lifting hooks in the core. These wires should be long enough to extend 3 or 4 inches above the top of the cope flask.

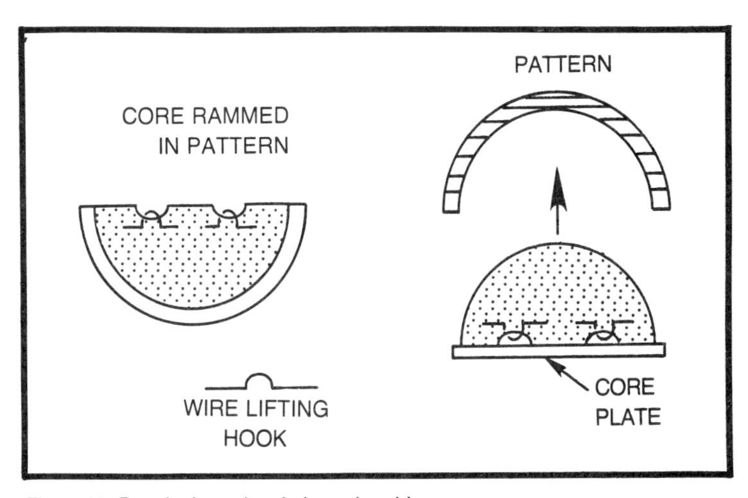

Fig. 7-43. Developing a bowl shaped mold.

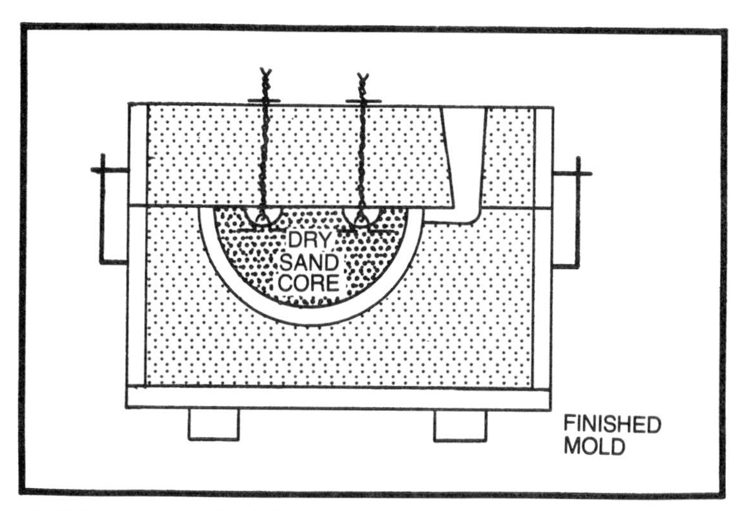

Fig. 7-44. Completed bowl shaped mold.

The cope is rammed in the usual manner, a bar is passed through the hanger wires, which are protruding through the cope. The wires are twisted together above the bar cinching or tying the core tightly against the cope face. The cope is lifted, carrying the dry sand core with it. The cope finished, the pattern rapped and drawn from the drag. The drag finished and the mold closed for pouring. See Fig. 7-44.

As you see bench molding presents many problems and many methods of solving these problems.

Chapter 8
Floor Molding

Floor molding is simply making molds that are too large to be produced on the bench. The operations are basically the same and performed in the same manner. The following are examples of floor molding.

Making A Large Gear Blank

A level bed of loose molding sand is fanned on the foundry floor, the cleats of the molding board rubbed in and the board leveled, the pattern placed on the board (cope face down) and the drag flask put into position, and the drag flask is clamped to the molding board in two or more places. Parting powder is applied lightly. The drag is riddled with molding sand through a #4 riddle covering the pattern and the molding board, with at least ½ inch of riddled sand. The drag flask is then filled to a depth of 5 inches with floor sand. The sand is now walked off. You get in the flask and walk down the sand with your feet. The sand is now peened around the inside perimeter of the flask and rammed with the butt end of your floor rammer. Add 5 inches walk down, peen and ram. This is repeated until the drag is completely rammed to a point 1 inch above flask edge.

The flask is then struck off with a strike off bar (a chunk of angle iron works well). The drag is well vented with a vent wire. The C-bar clamps are removed, floor sand is riddled on the drag using a #2 riddle to a depth of 1 inch and leveled. The bottom board is rubbed in, lifted off and the drag inspected for low spots which are corrected.

When the molder is sure that the bottom board is rubbed in correctly (the contact between the board and drag is 100% and level) the bottom board flask and molding board is clamped tightly together with C-bar clamps and wedges.

If the flask tends to move or slide during bottom board rubbing, the flask can be toenailed to the molding board. The rammed drag flask clamped between the molding board and the bottom board is now rolled over on to a riddled floor sand bed and leveled. You now check for rocking at each corner with your foot, and if any, wedge it out with wooden wedges. The clamps are removed. Remove molding board and set aside. The drag is blown off, checked and any defects taken care of.

As we are going to use a barred cope we must wash the bars and the inside of the flask with a clay wash. The clay wash is fire clay or bentonite mixed with water to a thick paint consistency. This helps hold the sand in the cope.

First we apply dry parting to the drag flask and riddle one inch of sand over the pattern and drag. We now brush the sand off of the flask joint with a molder's brush, so that when we place the cope on we will have a flask to flask joint with nothing between.

Now we clay wash the inside of the cope and its bars. Shake off excess wash before setting cope, (the bars have been cut back to conform with the pattern shape with sufficient clearance). The cope is placed on the drag. The gate stick and riser sticks are set into position along with any gaggers needed. Gaggers are L-shaped pieces of mild steel, or cast iron, used to support the sand in a deep pocket during drawing from the pattern, and when the mold is poured, reinforcing the sand. The gaggers are clay washed and pressed down into the sand with the shank against a bar. Another method used is "soldiers" which are square rough wood sticks which have no foot. These are clay washed. They need not touch the bars.

Now we clay wash the inside of the cope and its bars. Shake off excess wash before setting cope, (the bars have been cut back to conform with the pattern shape with sufficient clearance). The cope is placed on the drag. The gate stick and riser sticks are set into position along with any gaggers needed. Gaggers are L-shaped pieces of mild steel, or cast iron, used to support the sand in a deep pocket during drawing from the pattern, and when the mold is poured, reinforcing the sand. The gaggers are clay washed and pressed down into the sand with the shank against a bar. Another method used is "soldiers" which are square rough wood sticks which have no

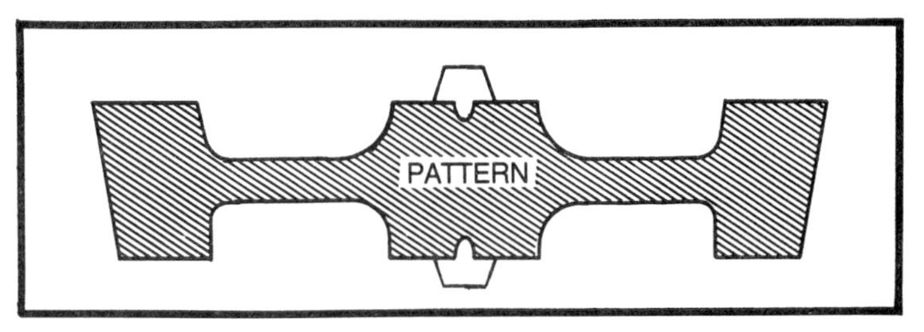

Fig. 8-1. Gear blank pattern.

foot. These are clay washed. They need not touch the bars but might simply be placed in the center of a deep pocket to assist with the lift (reinforcing the sand). Sand is shoveled into the cope approximately 5 inches, all gaggers, soldiers, sprue slicks and riser sticks that got knocked over are uprighted and tucked in place with your hands. The sand is tucked tightly under th bars with your fingers, placing one hand on each side of the bar and working down the length of the bar. The sand is now rammed as usual by peening and butt ramming, adding sand and repeating until the cope is complete. Do not strike the top of a bar when ramming as it will loosen the sand under the bar. The cope is struck off, the riser sticks and sprue pin removed, the pouring basin cut. The cope is lifted, not rolled over but set on saw horses as lifted. The cope is blown off from below and above. Any loose spots, small drops or damages are repaired. The pattern is swabbed, rapped and drawn, gates cut and finished, the drag blown out, core set and the mold closed, clamped and weighted for pouring. See Fig. 8-2.

Bedding In

Often times we have a large solid pattern which is best molded by placing the drag flask on a level sand bed, pins up, driving stakes against the flask to prevent it from moving. The flask is filled with loose molding sand and walked off, then covered with 1 inch of riddled sand, the pattern pressed down into the sand and driven down with a raw hide mallet. The sand is rammed and peened under the pattern working up to the parting line of the pattern and troweling the parting off tight and smooth. The pattern is removed to check the parting line and repair any soft or improperly rammed areas. Add sand to soft spots, bump pattern down, draw pattern and slick spots with spoon.

When the drag is finished, with the pattern in place, the drag is vented and parting powder shaken on the parting face. The cope is

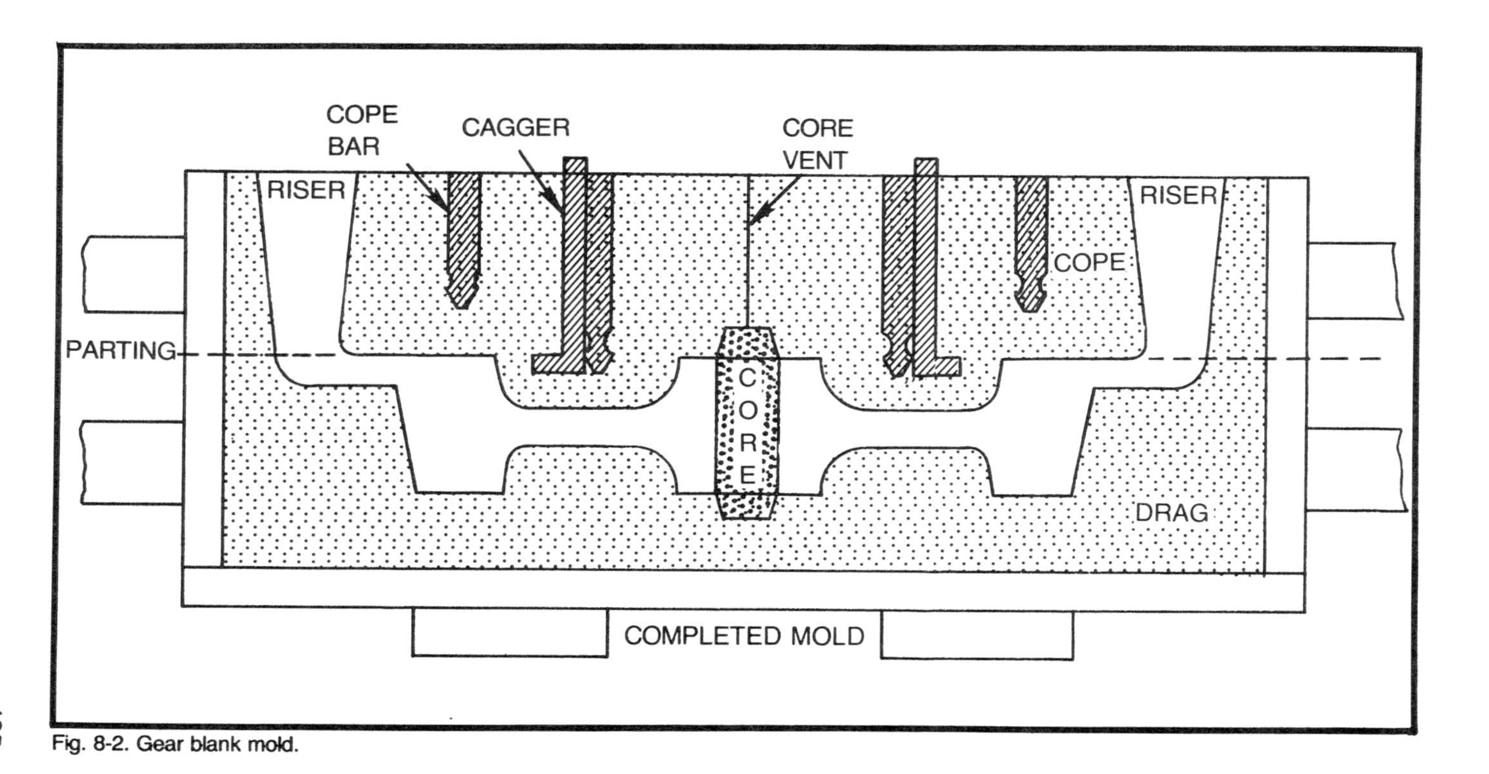

Fig. 8-2. Gear blank mold.

then placed on and rammed in the usual manner, lifted, the pattern removed. The mold is then finished and closed. See Fig. 8-3.

False Cope

If we have a case where the pattern could not be bedded and properly rammed to produce a good usable drag due to its shape, we make up what is called a false cope. In this case we place a cope parting face up on a bottom board. Fill this with floor sand, walk it off, place the pattern cope side down in the sand and ram the sand up to the parting line with no regard to how the sand is under the pattern. We are simply establishing a parting line and face upon which we can ram the drag correctly, as it could not be bedded in properly. The parting is troweled down and any sand removed from the flask edge.

The cope and pattern is then dusted with dry parting and the drag flask put on, the drag is rammed in the usual manner (venting etc.) and a bottom board rubbed on. The entire assembly is clamped together and rolled over on a sand bed. The clamps are removed, the cope and cope bottom board removed. This leaves the drag properly rammed and parted. The cope we used is shaken out, the empty cope is placed back on the drag and rammed as usual. What we have

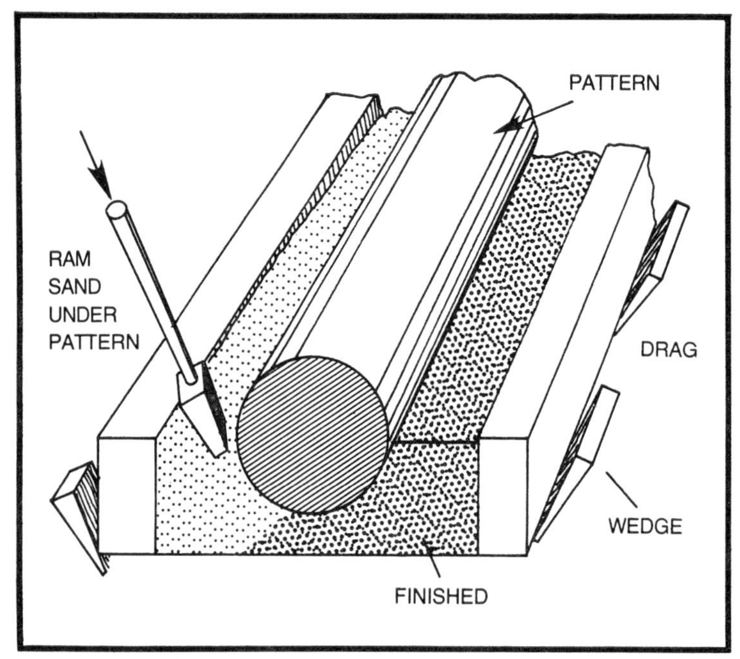

Fig. 8-3. Bedded in mold.

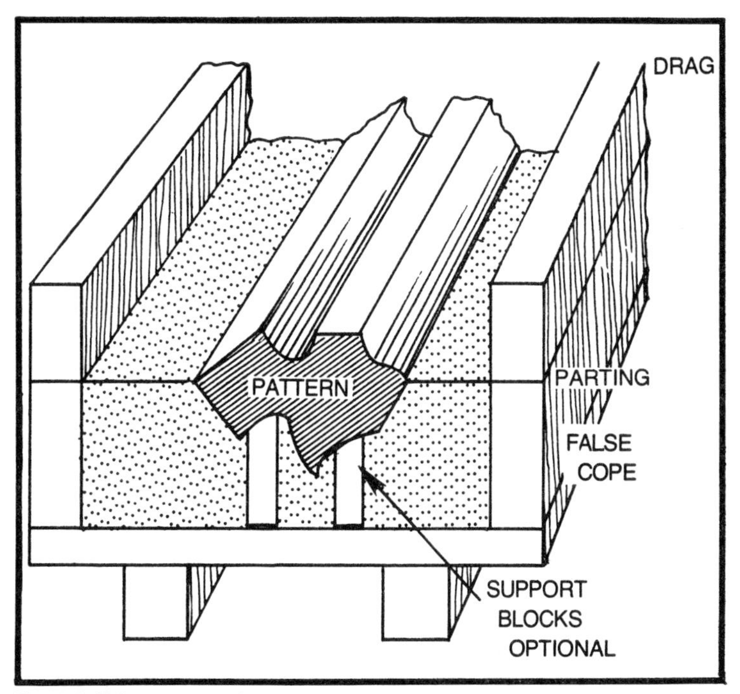

Fig. 8-4. False cope mold.

done is use a cope flask as a green sand follower, just to let us make the drag correctly. A false cope is some times called an odd side. See Fig. 8-4.

Roll-Off Cope

If the pattern is a flat back with nothing in the cope, or if the portion in the cope is heavily drafted or half round, the cope can be rolled off on roll off pins and guides, like opening a book. In this case the cope is not lifted but supported with a prop or two and finished in that position then closed like it was opened. See Fig. 8-5.

Open Sand Molding

Open sand molding, as simple as it looks, is extremely tricky to say the least. Two problems exist, ramming and pouring. Two straight edges are set up on bricks on the foundry floor and carefully leveled in all directions by using wedges. When level, molding sand is shoveled on each side of both straight edges and butted with a floor rammer just tight enough to prevent them from shifting or getting out of level. See Fig. 8-6A.

Now that we have our straight edges up and in level NES&W, check again and correct if necessary.

Now the space between the rails is filled with molding sand and walked down, (don't ram) and is again filled with loose molding sand to brimming full. Now with a straight edge strike the sand off level with the tops of your straight edge guides. Do this carefully with a seesawing action of the strike off, striking off only a portion at a time. Riddle 1 inch of sand as evenly as possible over the struck off bed, and clear the sand from the top of the straight edge guides. The riddled sand is now struck off level by putting a 1 inch thick block between the straight edge on each end. See Fig. 8-6B.

The actual ramming of the 1 inch level layer of riddled sand (on our level bed) is done by beating it down to a firm surface, level with the guide rails, commencing at one end, gradually advancing and bringing the straight edge down in a number of smart blows until the entire surface of riddled sand is level with the rails. The top is now slicked smooth with the finishing trowel, not tight, just smooth.

The diameter of the disc we are to cast is layed out on the troweled surface with a pair of trammels. A wood templet 1 inch thick is cut on a bandsaw to the diameter of the lay out on the sand.

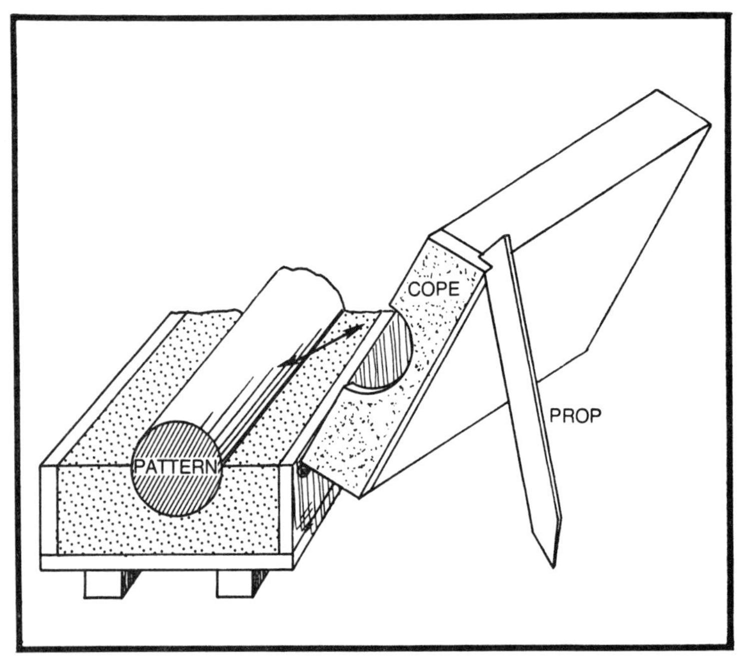

Fig. 8-5. Roll off cope.

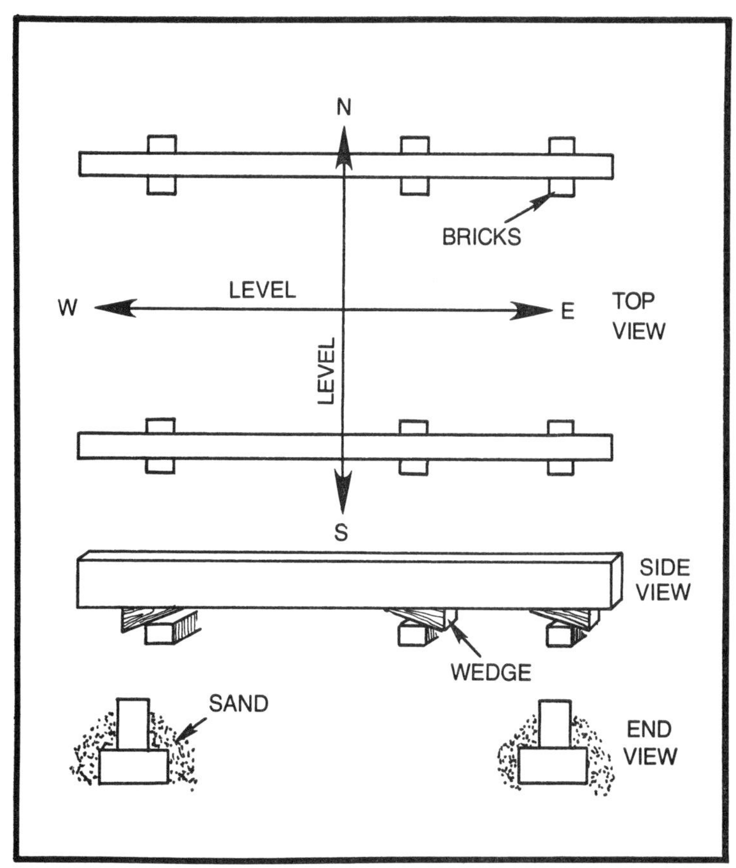

Fig. 8-6A. Open sand molding, step 1.

The templet is placed on the line and sand banked against its outer edge, lightly butted in place and troweled off level with the templet. This is continued until you have completed the circle. See Fig. 8-6C. Pig iron is layed against the sand to keep it in place. See Fig. 8-6D.

A pouring and runner basin is built on the top in an open ended frame. A dry sand core protects the edge of the runner basin and open mold cavity from washing away the green sand. See Fig. 8-6E.

The open mold is vented with a curved vent wire.

The casting face, in the drag, can be made smoother by dusting the cavity with graphite from a blacking bag and slicked down smooth with a brush and your fingers. The excess blown out. To prevent an over pour you can make the mold cavity 2 inches deep (for a 1 inch thick plate) and cut several flow off channels at the 1 inch depths to allow the excess pour flow off to the 1 inch level. See Fig. 8-6F.

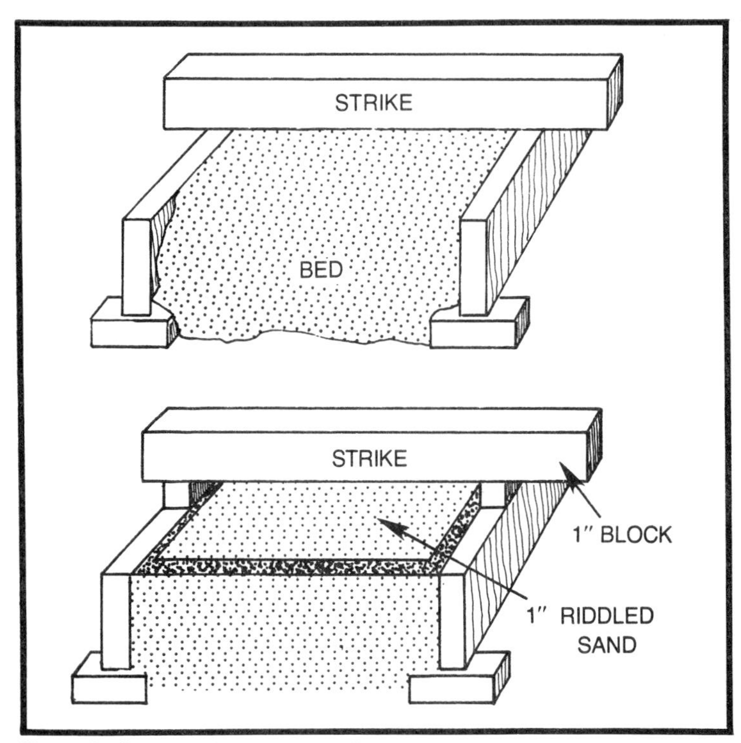

Fig. 8-6B. Open sand molding, step 2.

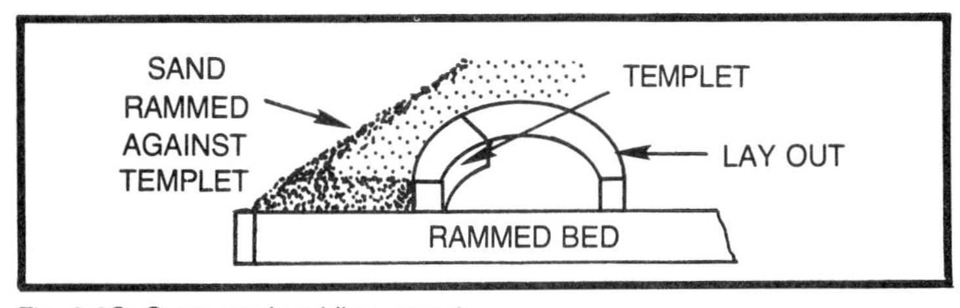

Fig. 8-6C. Open sand molding, step 3.

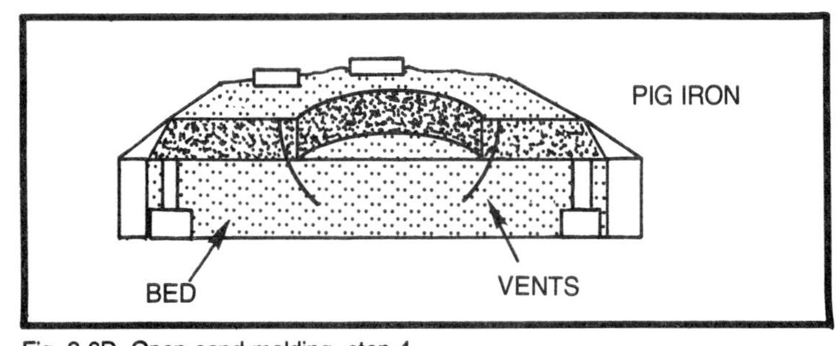

Fig. 8-6D. Open sand molding, step 4.

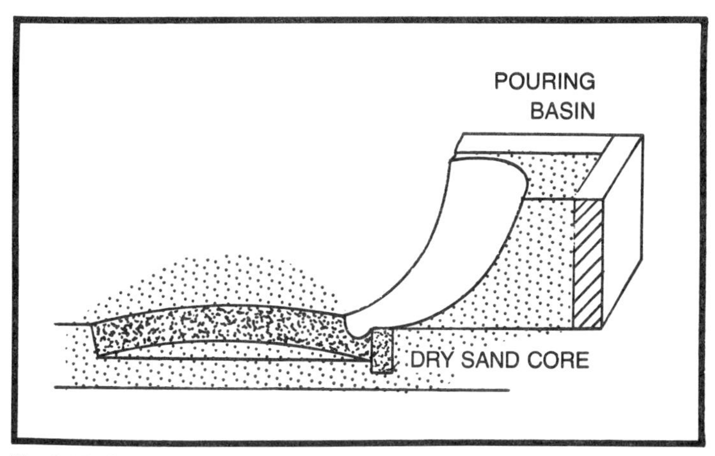

Fig. 8-6E. Open sand molding, step 5.

Draw Back

Often times you will have a portion of a pattern that cannot be drawn with the pattern but must be drawn into the cavity horizontally and then out. The large tapered hub on propellers usually have what is called chucking pads cast on the hub which must be straight sided. The pads on the casting are used to chuck the casting on a boring mill to bore the hub (they give the machinist a straight side to chuck). See Fig. 8-7A.

The hub is bored then the casting is mounted on an arbor through the bore and the chucking pads machined off. In molding the pads, patterns are pinned to the side of the hub with long wooden

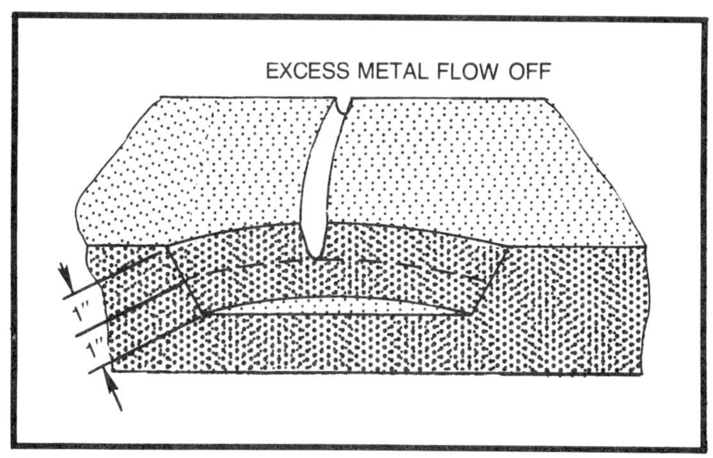

Fig. 8-6F. Open sand molding, step 6.

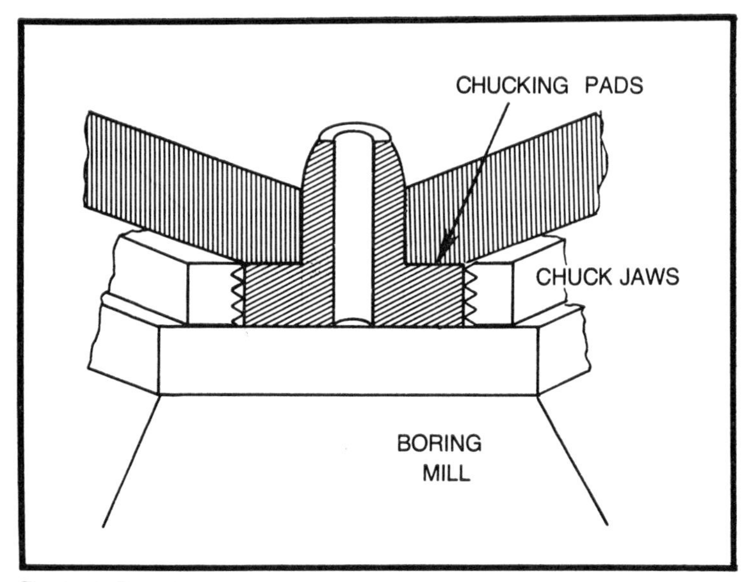

Fig. 8-7A. Draw back molding, step 1.

dowels which pass through the pads and into the pattern. These dowel pins are just snug enough to hold the pads in position. When ramming the job up, the sand is rammed around the pads so as to hold them in position and the pins are pulled out, freeing the pads from the pattern body. The ramming is completed. The pattern is drawn leaving the pads in the sand, the pads are then drawn in and out. See Fig. 8-7B.

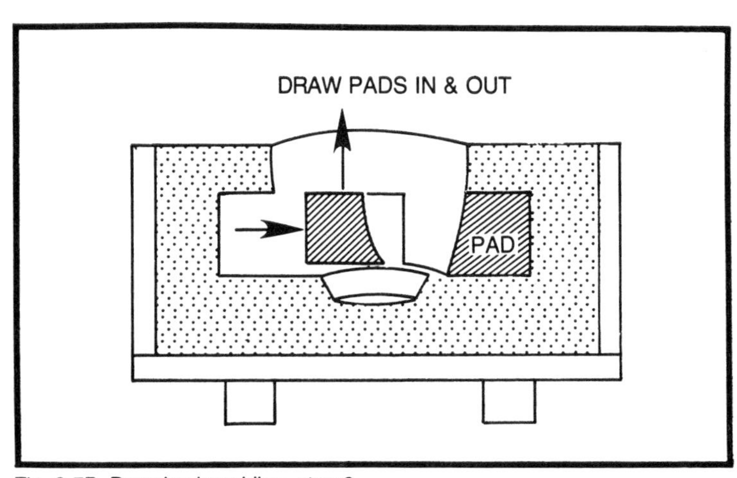

Fig. 8-7B. Draw back molding, step 2.

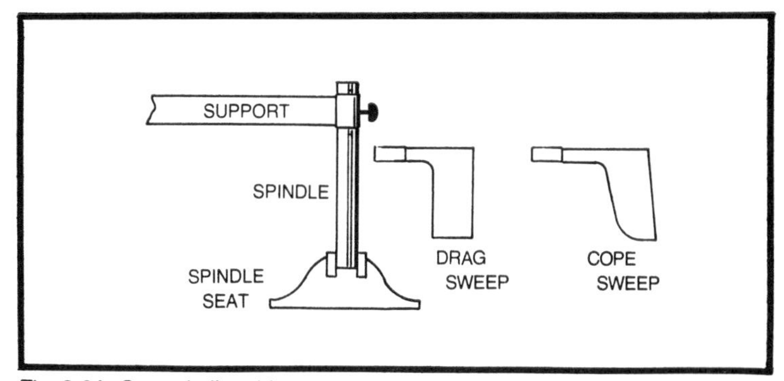

Fig. 8-8A. Swept bell mold, step 1.

Swept Bell

Here you must have a spindle seat, a spindle shaft, a shaft support and two sweeps, one for the cope and one for the drag. See Fig. 8-8A.

The spindle seat is leveled upon the floor in a drag flask. The spindle is placed in the spindle seat and plumbed true in all directions. The drag is rammed up and struck off level. The cope sweep put into position and the support attached. Punch vents into the drag around the spindle seat. See Fig. 8-8B.

You might, at this point, wonder why the cope sweep is used on the drag half of the mold. The cope sweep represents the shape we want in the cope and will clear up when we make the cope.

Sand is heaped up around the spindle, and the sweep is rotated as we ram the sand around the spindle. The inside face of the sweep scraping the sand to conform to its shape. We ram and sweep until we have on the drag a rammed and swept body of sand which represents the shape of the mold cavity we want in the cope. We

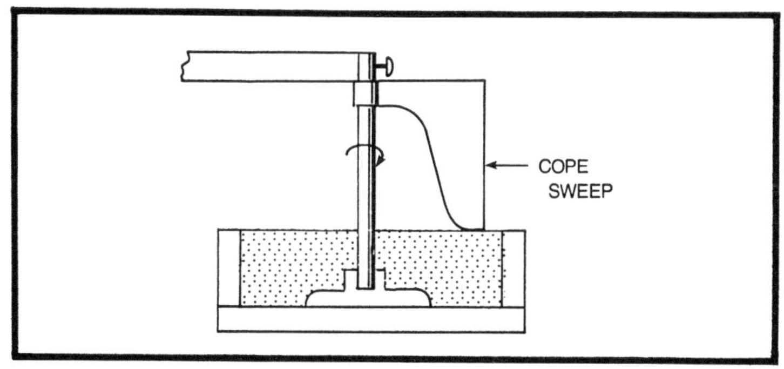

Fig. 8-8B. Swept bell mold, step 2.

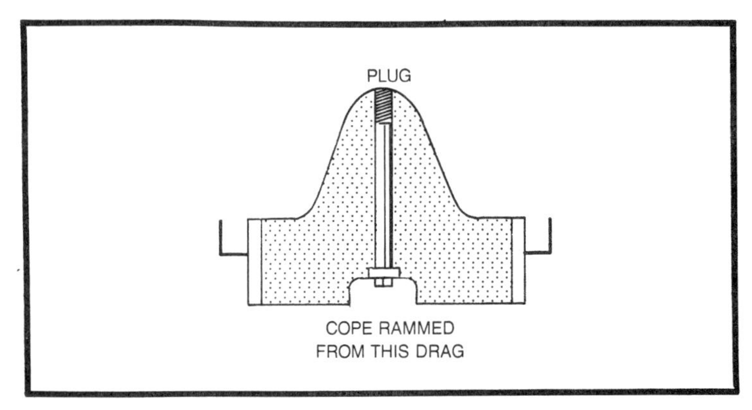

Fig. 8-8C. Swept bell mold, step 3.

now remove the support, cope sweep and spindle. The hole in the sand left by removing the spindle is plugged up with a tapered wooden plug. See Fig. 8-8C.

We dust the sand surface with parting powder and place the cope on the drag and ram it up in the usual manner (venting etc.). Lift the cope off and set it on horses and finish it. The cavity in the cope represents in reverse (female) the shape of the drag mold so if closed there would be no cavity left. Here's where the drag sweep comes in. We remove the wood plug, replace the spindle, place the drag sweep on the spindle and attach the support. The curve of the inner face of the drag sweep is closer to the center of the spindle than the cope sweep by the metal thickness of the bell casting we are making.

With this sweep we carefully sweep off the metal thickness, and remove the support sweep and spindle. The hole left by the spindle

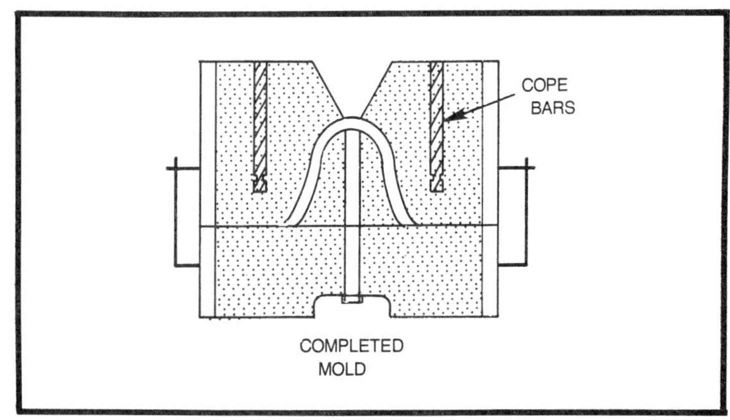

Fig. 8-8D. Swept bell mold, step 4.

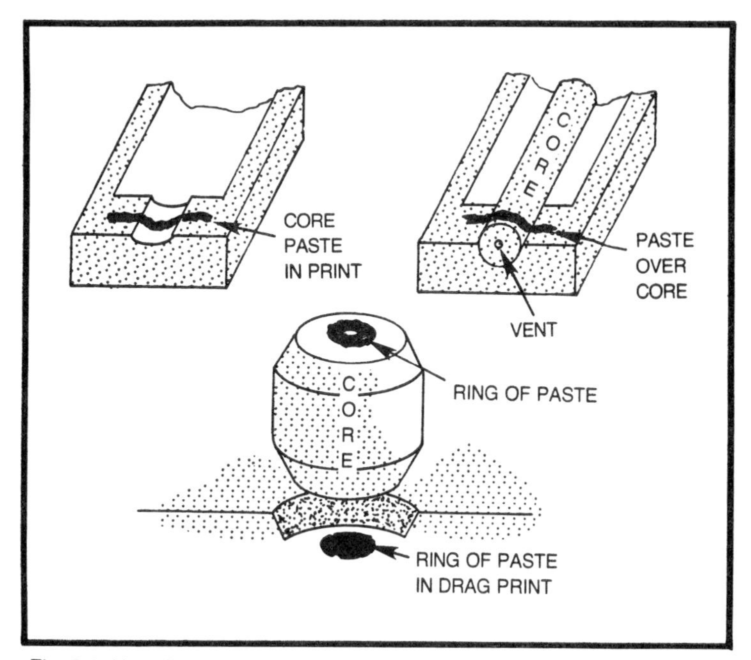

Fig. 8-9. Use of core paste.

is now closed up with a tapered dry sand core, first fill the hole up with molding sand. The drag is finished, blacked, slicked etc. The cope placed on the drag clamped and weighted, ready to pour. See Fig. 8-8D.

In both chapters on bench and floor molding we did not go into any detail as to the gates and risers required in the examples as we will cover them separately in Chapter 9. Because most castings can be poured by more than one way with success, the choice is yours. I have produced swept bells as in Fig. 8-8, gating some simply through the top and some through the lower rim.

The mold, bench or floor, can be green sand, skin dryed green sand, or dry sand. The weight of the casting and its complexity will dictate which way to go. Some shops pour 500 lb. castings in green sand with excellent results and other shops pour everything from 150 lbs and up in skin dryed or dry sand molds. To skin dry a mold, the cope and drag faces and mold cavities are sprayed with a mixture of molasses water, glutrin water (with or without graphite) and dried to a depth of ¼ to ⅜ of an inch back from the surface. This is done with a torch using a soft flame. The flame must be kept in continuous motion (sweeping) to avoid hot spots or burning the sand. All you want to do is dry and set the binder (glutrin etc.) and produce a dry firm skin on which the metal will lay. Prepared mold sprays and washes can be purchased. Some are mixed with alcohol, sprayed or swabbed on the sand and ignited, the alcohol burns and sets the wash. Some are universal, used for both a core and mold wash, or a spray. A good one is made up of zircon flour, graphite and a binder sold in a paste form. You supply the alcohol. The trade name is Zirc-O-Graph A.

A dry sand mold is one that is made up of sand mixed with a binder such as pitch, resin etc. (in steel flasks) and the cope and drag dried, in a core or mold drying oven, until dry throughout, when cooled are closed and poured. A skin dried mold must be poured soon after drying, otherwise the moisture, behind the skin dried depth, will migrate back to the surface and defeat the purpose. Plaques and grave markers are always skin dried to prevent the lettering from washing away when poured.

Core paste applied from a paste bulb is used as a seal to prevent metal from running over the section of a core in a print and getting into the core vent, closing it off. A line of core paste is run across the print in the drag, the core set, and a line run across the core and drag which will seal off the cope. A circle of core paste is applied to the bottom of a drag print, the core set and a circle of core paste applied to the top of the core to seal off the cope. See Fig. 8-9.

A ring of core paste is often placed on the drag around the cavity to seal cope and drag together as an added safety against run outs. In

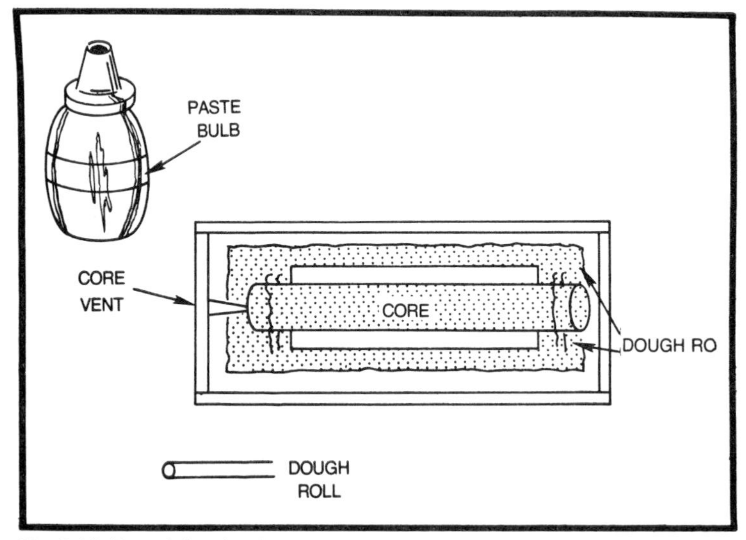

Fig. 8-10. Use of dough roll.

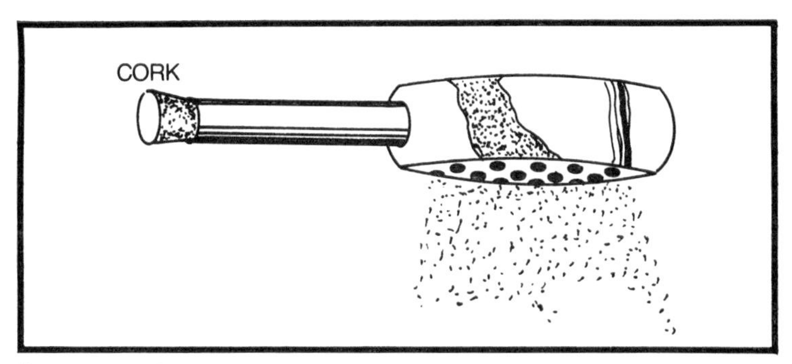

Fig. 8-11. A sifter for parting sand.

floor work what is called a dough roll is used. The dough roll is made of wheat flour and molasses water and rolled out into worms with the palms of the hands on a board, like you made clay worms as a child. See Fig. 8-10.

Do not close off core vent exit cut in the molds surface (this gas must get out). When the mold is closed the dough roll forms a good flat seal when it squishes flat.

Also in medium to large floor work, a dry sharp silica sand is used for parting in place of dry parting powder, applied with a parting sand sifter. See Fig. 8-11.

The bigger the core the bigger the vent should be.

Chapter 9
Gates and Risers

There is almost an endless variety of gates and risers and combinations of them. It is impossible to do anymore than cover the reasons we use certain combinations.

The basic parts of a gating system are the pouring basin, the sprue, the runner and the ingates. See Fig. 9-1.

The object of any gating system is to allow the mold to be filled as rapidly as possible with a minimum amount of turbulence and to provide sufficient hot metal to feed the casting during solidification to prevent shrinkage defects. As a casting will solidify from the thinnest section toward the heavier section and in doing so draws liquid metal from the heavy section, therefore the heavy section must be supplied with a riser or reservoir of hot metal, or it would shrink as it gives up its liquid metal to feed the solidifying thin section. Consequently, we must gate into the heavy section and provide it with feed metal. See Fig. 9-2.

The metal entering the riser must be hot (the last metal to enter during pouring) to promote feed metal based on directional solidification.

The gating system must be designed to trap any dirt or slag washed in during pouring. The most effective system is to place the runner in the cope and the gates in the drag, the gates angling back in the opposite direction of the flow of the metal in the runner.

Now the metal runs down to the end of the runner carrying any dirt, or slag past the gates, trapping it at the end of the runner, then

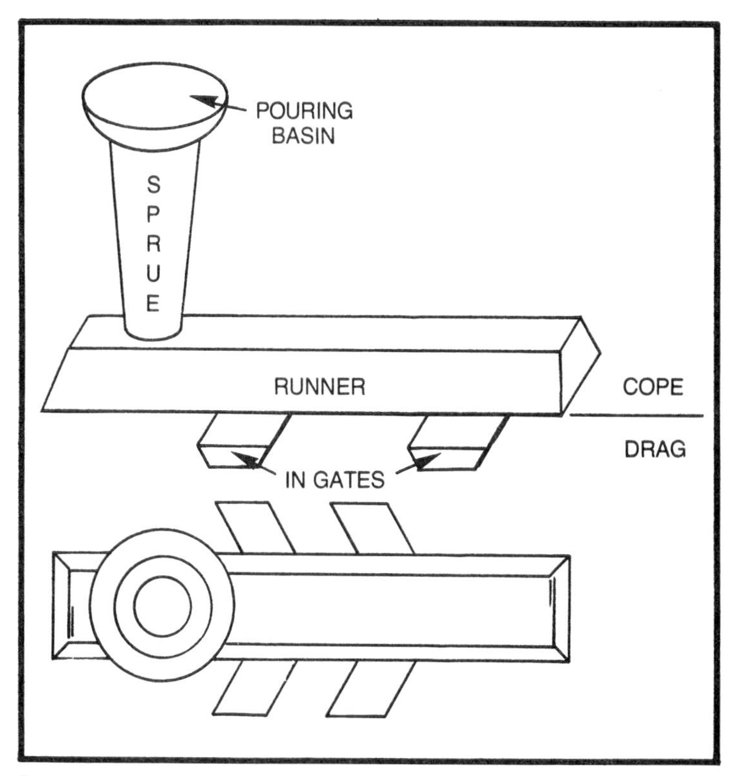

Fig. 9-1. Elements of a gating system.

clean metal backs up into the gates and runs into the casting. See Fig. 9-3.

The use of a runner in the cope with angled back drag gates is the most common type of gating on match plates with multiple patterns.

The runner bar is wide enough and extends into the cope to a point above the highest point of the castings. This way the runner becomes riser and runner combined. See Fig. 9-4.

Gating and risering is by far the most controversial subject in the foundry and each expert comes up with his pet theory on the subject.

I will attempt to give some general rules. However, each metal has its own peculiarities, shrinkage rates etc. These must be carefully studied in order to determine just where to start. Years ago in New Orleans, we had a very large aluminum gear housing to cast, 7 feet in diameter and quite deep. It was a new job from start to finish. We made the pattern and core boxes in our own pattern shop. Every

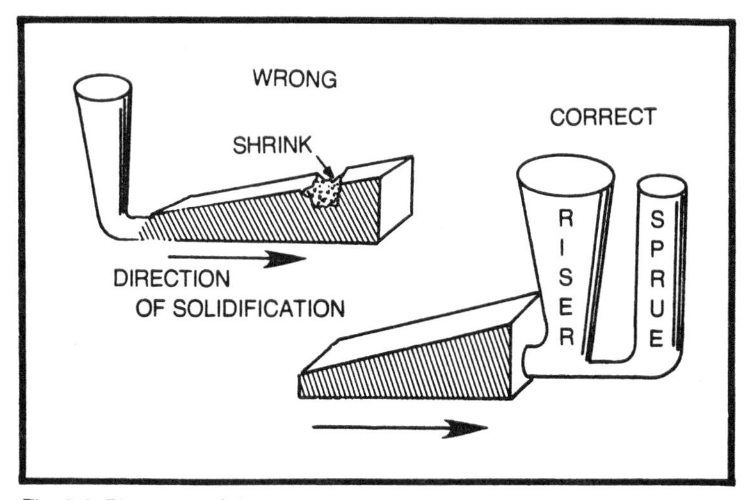

Fig. 9-2. Placement of risers.

molder in the shop had a different idea as to how it should be gated, each sure his was the correct way to go but all of them different.

As we had some 200 to produce I decided to use the first casting as a guinea pig. We made a mold and gated it into one side with a simple sprue and choke, knowing all too well we were undergated. We used no risers whatsoever. The casting ran and shrank, everywhere it needed a riser. By postmortem study we were able to determine exactly where to gate and what size, number and location the risers should be, which in this case did not agree with what we had thought before casting the beast. The job was a complete success from there on out.

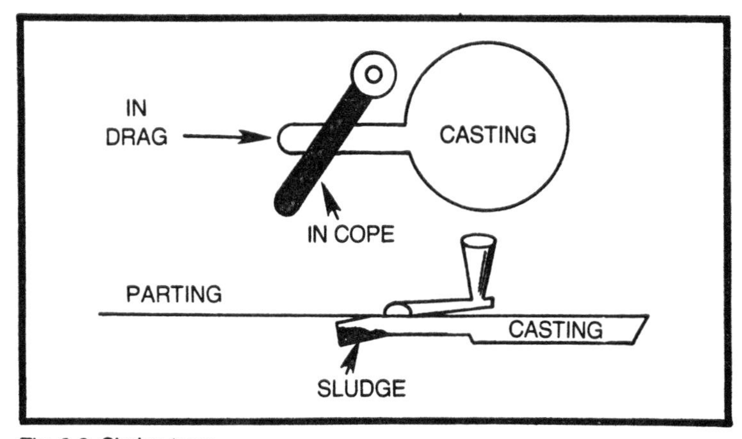

Fig. 9-3. Sludge traps.

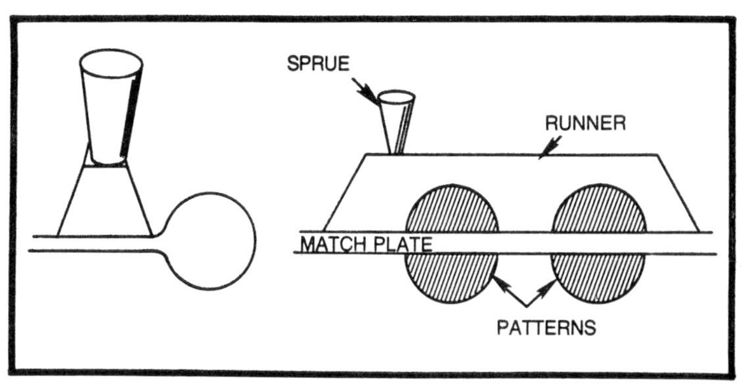

Fig. 9-4. Combined runner and riser.

Back to the general rules for gates and risers.

1. You must have directional solidification from the light sections to the heavy sections.
2. The pouring basin should be at least three times the diameter of your sprue.
3. The sprue should be small enough in diameter so that it can be kept choked during pouring.
4. The gating system should allow the metal to enter the mold fast but without turbulence.
5. Use sufficient number of gates properly placed to prevent cold shots and miss runs.
6. Design the gating system in a manner to prevent the metal from splashing through the gate.
7. Avoid any gate design that would spray the metal into the mold cavity.
8. Gate through the risers if possible, but when doing so, choke between riser and sprue.
9. Avoid gating against a core. Come in on a tangent so that the metal will not splash against the core.
10. Avoid all sharp corners in the gating system (use large fillets).
11. Patch and clean out gates, give them the same attention that you give the mold cavity.
12. Do not cut gates by hand. If possible, used mounted gates or set gates.
13. Use round down gates, cut with a tapered metal sprue cutter. When using a gate stick it should be smooth so that the downgate sand will be firmly rammed and smooth.

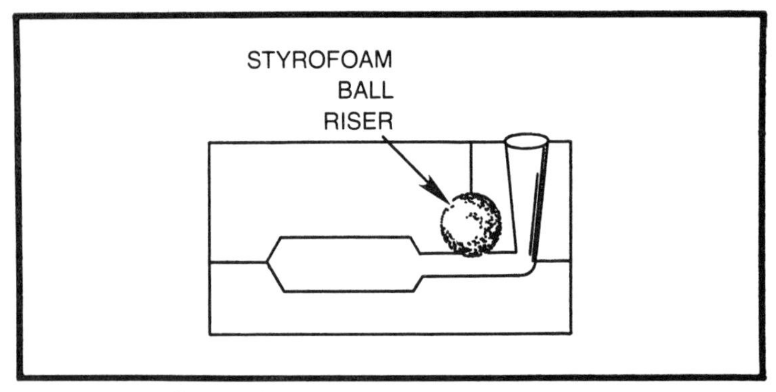

Fig. 9-5. A Styrofoam ball riser.

14. A piece of paper or tin should be placed over the pouring basin of closed molds until they are ready to pour.

A fairly new and very effective system of risering is the use of styrofoam balls, a sphere makes an excellent feed riser. The styrofoam ball is stuck down on a tapered gate stick and rammed in the cope. The metal during pouring vaporizes the ball and takes its place. See Fig. 9-5.

With small castings of a thin even wall thickness, the sprue will serve as both sprue and riser provided the ingate is not too thin that it will freeze off before the casting has solidified. See Fig. 9-6.

In going from a liquid to room temperature, you have three distinct shrinkages:

1. You have, as the metal cools from a molten mass to its solidification temperature, one reduction of volume.
2. When the metal goes from its solidification temperature to a solid you have another reduction in volume. It is during this second period that you must supply metal to compensate for the reduction in volume. This is why risers are used.

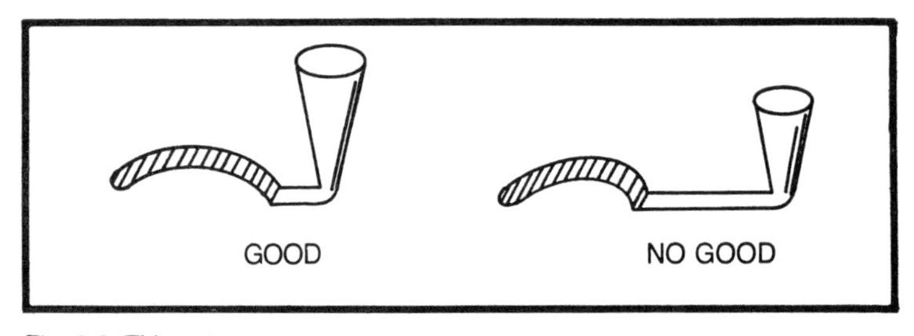

Fig. 9-6. Thin gates.

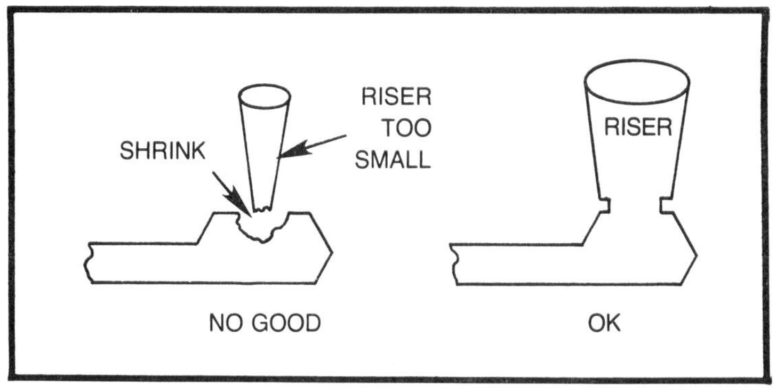

Fig. 9-7. Small risers.

3. From the solid stage to room temperature there is another reduction in volume. This reduction is taken care of by the use of a shrink ruler when the pattern is designed.

It should be pointed out that an ineffective riser, one that will not feed is worse than none at all. See Fig. 9-7.

When you have an area or section on a casting that requires feed metal, but by virtue of your design and the location in the mold, a feed riser cannot be placed where needed, a chill is used. An external chill made of cast iron molded in the sand will chill the area (where applied) and promote directional solidification.

An internal chill is one that (like a chaplet) becomes part of the casting fusing into the area of use. See Fig. 9-8.

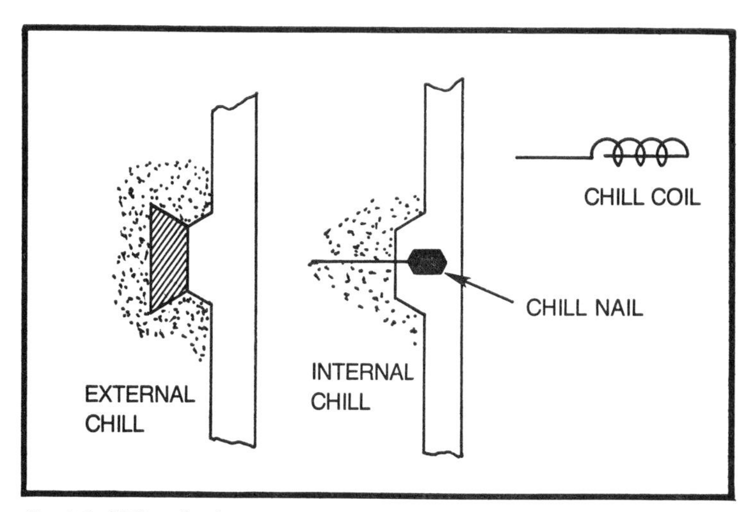

Fig. 9-8. Chill methods.

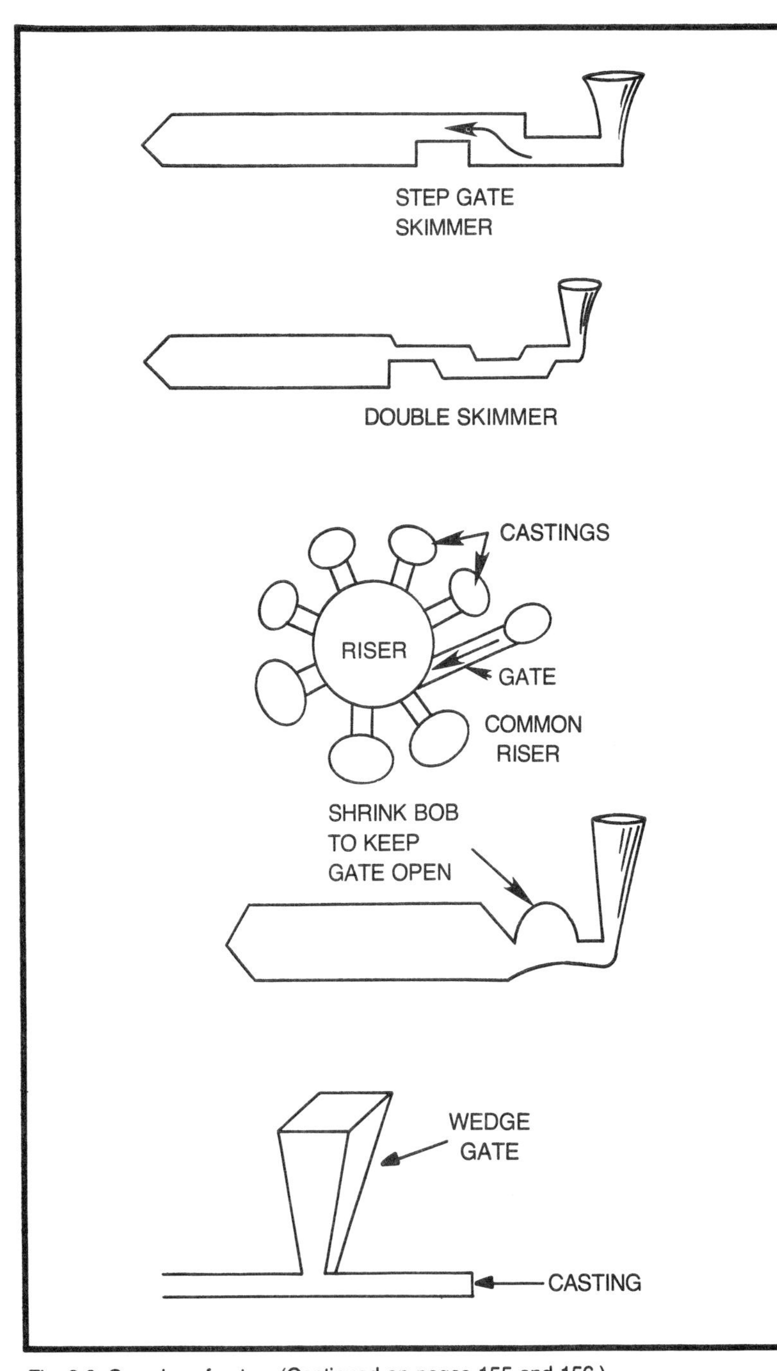

Fig. 9-9. Samples of gates. (Continued on pages 155 and 156.)

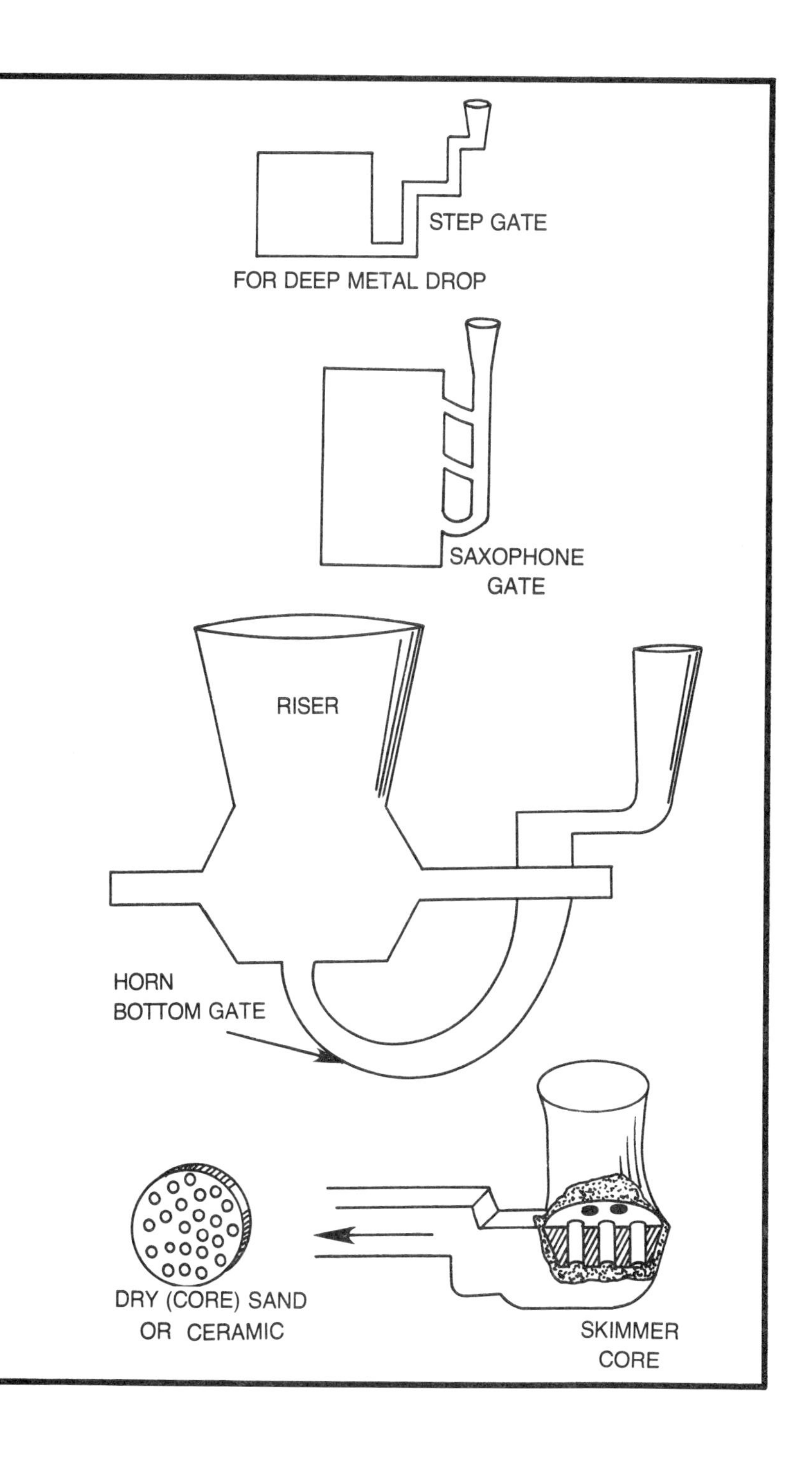
STEP GATE
FOR DEEP METAL DROP
SAXOPHONE
GATE
RISER
HORN
BOTTOM GATE
DRY (CORE) SAND
OR CERAMIC
SKIMMER
CORE

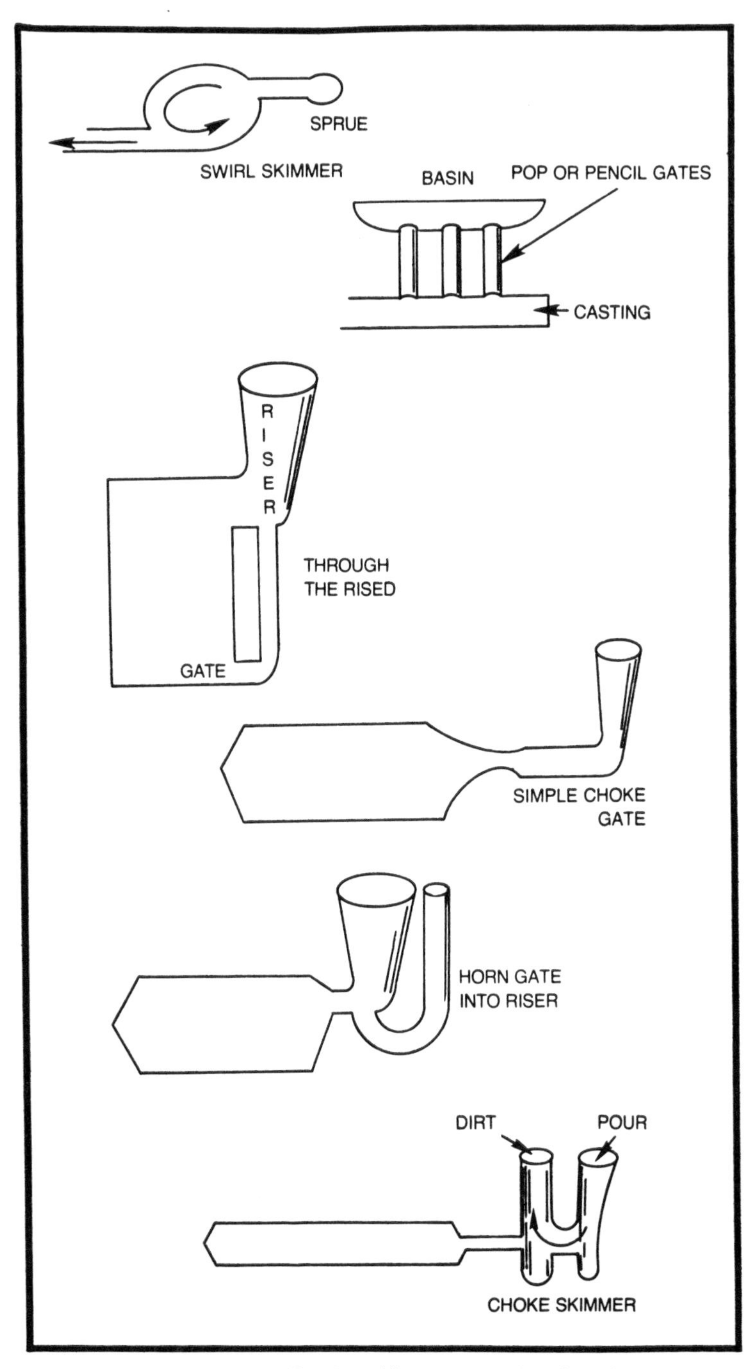

Fig. 9-9. Samples of gates. (Continued from page 154 and 155.)

Box car wheels are chilled with a ring chill to make the tire dense and hard (for wear) leaving the web soft and resilient. The face of an external chill which will be in contact with the casting is washed with lube oil and graphite to prevent blowing. A variety of other gate and riser designs are shown in Fig. 9-9 with a description of their uses.

Chapter 10
Designing Sand Castings

More often than not a casting is designed by someone who doesn't have the foggiest idea of how or why the casting is to be produced and the foundry will inherit problems when accepting such a poor design from the casting standpoint. Someone who has no knowledge of the foundry and its requirements, should talk with the foundry before putting one line on the drawing board. We had such a case in New Orleans, La. We received a pattern and core box for a small 2 cycle air cooled engine cylinder. The bird who designed the cylinder had no knowledge of the process involved to produce the cylinder nor did he concern himself with this aspect. His main consideration was cosmetic and the ability of the fins to cool the cylinder. He designed the fins with sufficient surface, number, location and depth to cool the cylinder to the desired running temperature. Then, when he had arrived at this point, he added a 500 percent safety factor. This gave him closer and deeper fins. The company sent the prints out to a pattern shop. It should have been stopped at that point. Any pattern maker in his right mind could see from the print that if he made the pattern like the print, in no way could it be cast in green sand or any other sand. The print clearly read "sand cast gray iron class 30." Well, he made the pattern and core box, I guess he needed the money. When we got the pattern and core box, one look told us that in no way could it be cast. The fins were so close together and deep that the space between them could not be rammed, and if so, you could not draw the pattern and leave paper thin sand sections standing in the mold. Knowing the designer and how hard headed he was (after all this was his creation, his child), we decided, based on experience with other air cooled engine cylinders, that if we removed every 2nd fin on the pattern we could cast the cylinders easily and the engine would not suffer from insufficient cooling. As we needed the job, this is what we did. Well, after casting them for some five years, the company called the pattern in for redesign. We asked why the redesign and were told it was to increase the bore and decrease the finning, as the engine ran a shade too cool for proper operation. If they had noticed that the cylinder castings and pattern had only half the number of fins as the print called for, nothing was said, and we made the redesign patterns and kept on casting for the company. The redesign prints were correct for good sand casting practice.

There are lots of ways to skin a cat but only one to the cat.

BASIC DESIGN CONSIDERATIONS

1. On castings that must present a cosmetic appearance along with function, the exterior should be designed to follow simple flowing lines with a minimum of projections and no unnecessary projections.
2. The size and weight should be kept within the limits of adequate performance and the end use of the casting, not 1 inch thick sections when ¼ inch would do. Build in strength by design instead of weight or section thickness where possible.
3. Design where possible, so that all operations from concept to the end product are simplified; pattern equipment, core making, castings, machining etc. By doing this you will save time and money.
4. Avoid irregular or complicated parting lines, wherever possible. Design for partings to be in one plane.
5. Stick with designs that do not require loose pieces and use loose pieces only when it cannot be avoided, when there is no other way to accomplish the desired results.
6. Use ample draft but not excessive. Avoid any no draft vertical surfaces unless there is no other way out.
7. Design around dry sand cores. If you can have the casting leave its own green sand core, this is cheaper.
8. In designing a casting which requires a dry sand core, see that it has sufficient prints to support it without the use of

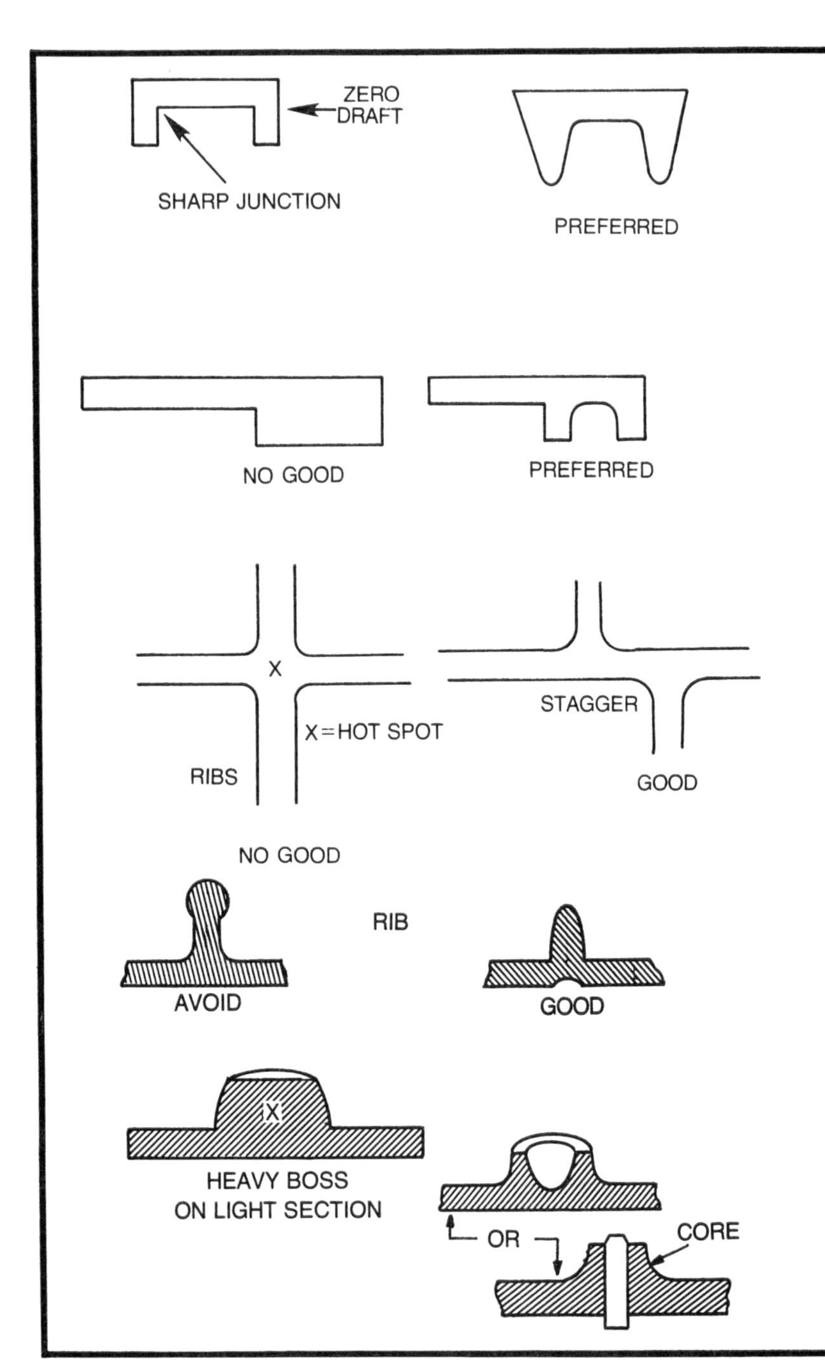

Fig. 10-1. Examples of good and bad casting designs. (Continued on pages 161 and 162.)

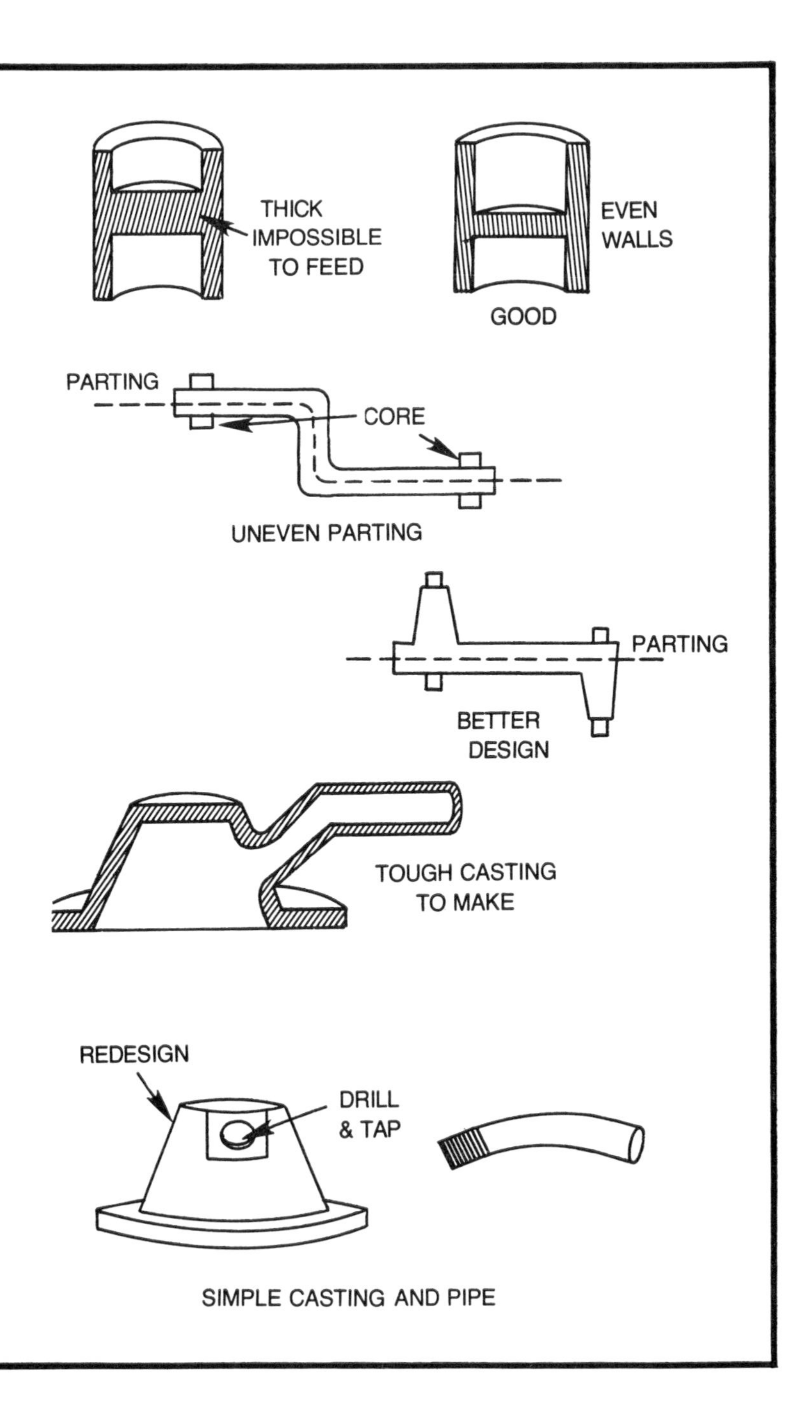
THICK
IMPOSSIBLE
TO FEED
EVEN
WALLS
GOOD
PARTING
CORE
UNEVEN PARTING
PARTING
BETTER
DESIGN
TOUGH CASTING
TO MAKE
REDESIGN
DRILL
& TAP
SIMPLE CASTING AND PIPE

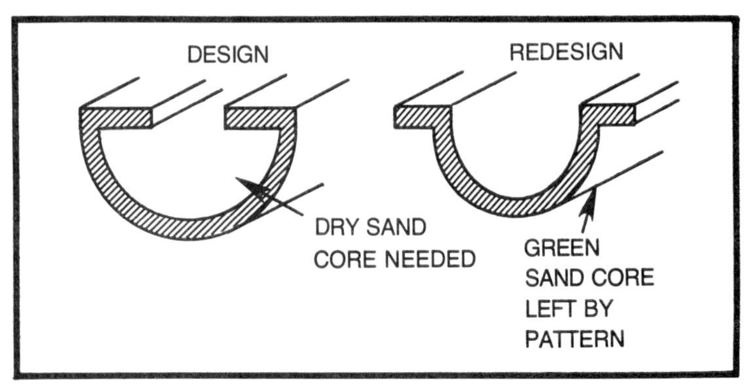

Fig. 10-1. Examples of good and bad casting designs. (Continued from pages 160 and 161.)

chaplets and that the print openings are sufficiently large enough to easily remove the core. (Core knock out.)

9. Avoid long slender cores through heavy metal sections or long spans. When unavoidable, they should be straight and well anchored to the mold.
10. Avoid metal inserts unless the section containing the insert is heavy enough to insure good purchase or the advantage gained by the use of inserts outweighs alternate steps.
11. Avoid the use of pattern letters on any surface other than one parallel to the parting.
12. Avoid sudden changes in section thicknesses creating hot spots.
13. Use ribs to stiffen or strengthen castings thus reducing weight.
14. Do not design a complicated casting in one piece, which would be much better produced as two or more castings welded or bolted together to give the desired results.
15. Do not try to cast sheet metal. Make a casting when the design dictates a casting, and go sheet metal when the design dictates it should be sheet metal.
16. Avoid all sharp corners and fillet all junctions.
17. Stagger crossing ribs so that the junction will not form a hot spot which could shrink.
18. If you can go any other way avoid deep pockets. If you cannot, call for cover cores in place of green sand cores.
19. Design so that the casting can be easily poured and risers are located at the highest point in the mold for easy removal.

20. Avoid designing heavy sections which fall below light sections in the mold that cannot be reached with risers. See Fig. 10-1.

ALTERNATE METHODS

Before designing an item or part to be sand cast, you should examine (if in your mind only) the possibility and feasability of producing it by any one of the many alternate methods of manufacture or fabrication, taking everything into consideration including cost. In other words you must make the decision based on many factors: Cost, quantity, weight, end use, product life, material, public acceptance etc. etc. (A broom should be made of straw not cast iron). There are many many items on the market today that are made of weldments, that should be castings and visa versa.

Ask yourself, sand casting or one of the following methods?

Die casting
Permanent mold
Investment casting
Shell (investment casting)
Screw machine
Stamping
Weldment
Forging
Extrusion
Header (hot or cold)
Four slide
Plastic injection
EDM
Sintered compact

If the answer is sand casting, you must then decide, based on metal, size, quantity, end use, cost etc., which way to go, green sand, dry sand, core mold, etc.

The point I am trying to make is you don't just up and design a sand casting. There are many other things which should directly influence your design. They must all be taken into consideration. In one large (one of the Big Four) auto manufacturing plants, where I worked at one time, processing castings, we would make an isometric drawing of the item or part, along with a positive of the interior and try in our mind to envision the patterns, core boxes, parting planes, flask equipment, gating, mold interior, casting, machining and what have you before getting down to the actual designing. If it was not suited to sand casting we shipped it off to the division (forging etc.) where we thought it should be.

We have given the basic rules for sand casting design. If you cannot incorporate these rules with any degree of satisfaction into an intended casting design, perhaps what you have or want to make is not a casting.

Chapter 11
Casting Defects

A knowledge of casting defects is essential. If you cannot pinpoint the cause of a defect there is no way of correcting the problem. Some defects are quite obvious, along with the cause. Some types of defects can often resemble each other in appearance and separating them is often difficult. A drawing or photograph of a defect is one thing and looking at the actual defect is another. You learn about defects by analyzing your own and those of others.

Here is a list of defects and their causes that should include anything you might ever encounter:

1. **Poured short:** Casting incomplete due to not filling the mold. This is a stupid trick which we all do—cause insufficient metal in the ladle. Going back and touching up the mold will not do it.
2. **Bobble:** Here the casting will resemble a cold shut or that it was poured short. The problem was a slacked or interrupted pour. When pouring a casting the metal must be poured at a constant choked velocity. If you slack off or reduce the velocity, you can cause this defect. A completely interrupted pour, start—stop—start, if only for a second. A great percentage of lost castings is caused by pouring improperly.
3. **Slag inclusion:** Slag on the face of the casting and usually down the sides of the sprue. The cause is not skimming the

ladle properly, not choking the sprue (keeping it brimming full from start to finish), sprue too big (cannot be kept choked) and gating system improperly choked.

4. **Steam gas porosity:** This defect usually shows up as round holes like swiss cheese, just under the cope surface of the casting and comes to light during machining. The cause is a wet ladle, where the ladle lining was not properly and thoroughly dried.
   In extreme cases the metal will kick and boil in the ladle. The practice of pigging the metal in a wet ladle and refilling it with the hopes that the pigged metal finished drying the ladel is sheer folly.
5. **Kish:** (cast iron) If the carbon equivalent of the iron is too high for the section poured and its cooling rate is slow, free graphite will form on the cope surface in black shiny flakes free from the casting, causing rough holey defects usually widespread. Carbon equivalent is the relationship of the total carbon, to the silicon and phosphorous content of the iron, which is controlled by the make up charge and silicon added at the spout. It is called carbon equivalent because the addition of silicon or phosphorus is only one third as effective as carbon, therefore the carbon equivalent of the three additives is equal to the total percentage of pure carbon plus one third the percentage of the silicon and phosphorus combined. The carbon equivalent is varied by the foundryman depending on what type of iron he is producing and what he is pouring.
6. **Inverse Chill:** This defect is found in gray iron castings and is hard or chilled iron in the center sandwiched between soft iron. The cause again could be incorrect carbon equivalent for the job or the presence of non-ferrous metals in the charge, lead, antimony, tellurium which are detrimental impurities.
7. **Broken Casting:** This could be caused by improper design. Improper filleting along with improper handling any where along the line. Copper base castings, red brass, yellow brass etc., are what are known as hot short, will break easily when hot. Thus, if the casting is shaken out of the mold before it has cooled sufficiently, it can get broken very easily. A mold or core which has too high a hot strength will not give or collapse to give the casting room to move as it shrinks and will break the casting.

8. **Bleeder:** This defect is caused by shaking the casting out too soon when a portion of it is still liquid. The section runs out leaving a defect. In extreme cases the entire center section will run out on the shake out man's feet.
9. **Lead Sweat:** A covering of lead tears on the outside of a high leaded bronze or brass casting with underlying holes or porosity (red metals containing a large percentage of lead such as leaded bearing bronze). Since the lead has the lowest melting point of the constituents of high leaded bronze (copper 70 percent—tin 5 percent—lead 25 percent) and is not in solution with the copper and tin, it remains liquid until the casting has cooled below 620 degrees Fahrenheit. If the casting is shaken out before it is below this temperature the lead will sweat out from between the copper and tin crystals.
10. **Run Out:** This is caused by the metal in the mold running out between the joint of the flask, which drains the liquid metal partially or completely from any portion of the casting above the parting line. A run out can come through or between the drag and the bottom board from a cracked drag mold. Also, it can come from between a loose improperly fitting core and the core print. It is caused by insufficient room between the flask and the cavity, insufficient weight on the cope (cope raises during pouring) or improperly clamped molds. Excessive hydrostatic pressure (sprue too tall for the job) no dough roll used between cope and drag (large jobs) or a combination of all of the above.

    An attempt to save a run out by placing your foot on top of the cope and applying pressure above the point. where the liquid metal is running out is foolhardy and dangerous. Also trying to stop off the run out flow with sand or clay is folly.
11. **Omission** of a core: Results and cause obvious.
12. **Ram Off:** Is a defect resulting from a section of the mold being forced away from the pattern by ramming sand after it has conformed to the pattern contour. This is caused by careless ramming where the mold is rammed vertically and then on an angle, causing the vertically rammed sand to slide sideways leaving a gap between the pattern and the sand, resulting in a deformed casting. Another cause is using a sand with poor or low flowability. (Too much clay or sand too fine.)

13. **Core Rise:** This defect is caused by a core rising from its intended position toward the cope surface, causing a variation in wall thickness or if touching the cope no metal at that point. The core has shifted from its position. A green sand core rise is when a green sand core in the drag is cracked at its base (caused when drawing the pattern) and floats toward the cope. Dry sand cores will float if the unsupported span of a thin insufficiently rodded core is too great, it will bend upward by the buoyancy of the metal. Insufficient core prints in number and design, insufficient chaplets, slipped chaplet, chaplets left out by molder, poor design of the core. This defect is easy to spot and remedy.
14. **Shifts:** These come in two classifications, mold shift and core shift.

    A mold shift is when the parting lines are not matched when the mold is closed, resulting in a casting offset or mismatched at the parting. The causes are: Excessive rapping of a loose pattern, reversing the cope on the drag, too loose a fit of the pattern pins and dowels, faulty mismatched flasks, too much play between pins and guides, faulty clamping, improper fitting (racked) jackets, and improper placing of jackets.

    A core shift is caused by not aligning the halves of glued cores true and proper when assembling them.
15. **Swell:** In this defect you have a casting which is deformed due to the pressure of the metal moving or displacing the sand. It is usually caused by a soft spot or too soft a mold.
16. **Sag:** A decrease in metal section due to a core or the cope sagging. The cause is insufficient cope bars, too small a flask for the job, insufficient cope depth. This defect will also cause misruns.
17. **Fin:** This defect is a fin of metal on the casting caused by a crack in the cope or drag, and is caused by wracked flasks, bad jackets, (and setting) uneven warped bottom boards, uneven strike off, insufficient cope or drag depth, bottom board not properly rubbed on drag mold sitting uneven (rocking).
18. **Fushion:** This defect is a rough glassy surface of fused melted sand on the casting surface either on the outside or on a cored surface. The cause is too low a sintering or

melting point of the sand or core. This is quite common when a small diameter core runs through an exceptionally heavy section (of great heat) which actually melts the core. This can also be caused by pouring much too hot (hotter than necessary) for the sand or cores. A mold or core wash can prevent this in some cases but if the sintering point of your sand is too low for the class of work, you need a more refractory sand, zircon etc.

19. **Metal Penetration:** This defect should not be confused with an expansion scab which is attached to the casting by a thin vein of metal.

    The defect is a rough unsightly mixture of sand and metal caused by the metal penetrating into the mold wall or the surfaces of a core (not fused).

    The defect is basically caused by too soft and uneven ramming of the mold or core, making the sand too (open) porous, also too high a pouring temperature, too sharp a corner (insufficient filleting) making it impossible to ram the sand tight enough, localized overheating of the sand due to poor gating practice, or molding with a sand too open for the job.

    Penetration in brass castings is sometimes traced to excessive phosphorus used in de-oxidizing the metal making it excessively fluid.

20. **Rough Surface:** This defect can run from mild penetration to spotty rough spots or a completely rough casting. Many factors or combinations are at fault—sand too coarse for the weight and pouring temperature of the casting, improperly applied or insufficient mold coating or core coating, faulty finishing, excessive use of parting compound (dust) hand cut gates not firm or cleaned out, dirty pattern, sand not riddled when necessary, excessive or too coarse of sea coal in the sand, permeability too high for class of work, core or mold wash faulty (poor composition).

21. **Blows:** Round to elongated holes caused by the generation or accumulation of entrapped gas or air.

    The usual cause of blow holes is sand rammed too hard (decreasing the permeability), permeability too low for the job, sand and core too wet (excessive moisture), insufficient or closed off core vent, green core not properly dried, incompletely dried core or mold wash, insufficient mold venting, insufficient hydrostatic pressure (cope too short), cope bars too close to mold cavity, wet gagger or soldier too close to mold cavity, poor grain distribution. Any combination of hot and cold materials which would lead to condensation as a hot core set in a cold mold or visa versa. A cold chill or hot chill will cause blows, also wet or rusty chaplets.

22. **Blister:** A shallow blow covered over with a thin film of metal.
23. **Pin Holes:** Surface pitted with pin holes which may also be an indicator of subsurface blow holes.
24. **Shrink Cavity & Shrink Depression:** These defects are caused by lack of feed metal causing a depression on the surface of the casting, a concave surface. The shrink cavity, a cavity below the surface but connected to the surface with a dendrite crystal structure.
25. **Cold Shot:** Where two streams of metal in a mold coming together fail to weld together. This defect is usually caused by too cold a metal poured too slowly or gating system improperly designed so that the mold cannot be filled fast enough.
26. **Misrun:** A portion of the casting fails to run due to cold metal, slow pouring, insufficient hydrostatic pressure, sluggish metal (non fluid due to badly gassed or oxidized metal).
27. **Scab:** Rough thin scabs of metal attached to the casting by a thin vein separated from the casting by a thin layer of sand. Usually found on flat surfaces, caused by hard ramming, low permeability and insufficient hot strength. Sand does not have enough cushion material, wood flour etc., to allow it to expand when heated. Unable to expand it will buckle causing scabs along with, but not always rattails, grooves under the scabs also called pull downs. It is the pull downs that bring about the scab.
28. **Blacking or Mold Wash Scab:** This is a case when the blacking or wash on the mold or core, when heated, breaks away and lifts off of the surface like a leaf and is retained in or on the metal. The cause is a poor binder in the wash, improperly dried wash or poor wash formula or all of the above.

29. **Sticker:** This is a lump or rat (bump) on the surface of the casting caused by a portion of the mold face sticking to the pattern and being removed with the pattern. This problem is caused by poorly cleaned, shellacked, polished pattern, rough pattern, cheap shellac, tacky shellac, sticky liquid parting, cold pattern against hot sand, insufficient draft.
30. **Crush:** This defect is caused by the actual crushing of the mold causing indentations in the casting surface. This is caused by flask equipment, such as bottom boards or cores that are too tall or too large for the prints or jackets. Also rough handling.
31. **Hot Tears:** This defect is actually a tear or separation fracture due to the physical restriction of the mold and or the core upon the shrinking casting. The biggest cause is too high a hot strength of the core or molding sand. These defects can be external or internal, a core that is overly reinforced with rods or an arbor will not collapse.

    If you restrict the movement of the casting during its shrinking from solidification to room temperature, it will literally tear itself apart.
32. **Gas porosity:** This defect is widely dispersed bright round holes which appear on fractured and machined surfaces. This defect is caused by gasses being absorbed in the metal during melting. This gas is released during solidification of the casting. Cause is poor melting practice (oxidizing conditions) and poor de-oxidizing practice.
33. **Zinc Tracks:** This defect is found on the cope surface of high zinc alloy castings. The defects are caused by the zinc distilling out of the metal during pouring. This zinc oxide floats up to the cope and forms worm track lines on the casting when the metal sets against the cope. The problem is caused by pouring too hot (metal flaring) in ladle or crucible. Pouring the mold too slow, insufficient gates. The mold must be filled quickly before the damage can be done.
34. **Drops:** This is where a portion of the cope sand drops into the mold cavity before or during pouring. The causes are bumping with weights, rough clamping, weak molding sand (low green strength) rough closing, jackets placed on roughly etc.
35. **Washing & Erosion:** The sand is eroded and washed around in the mold, some of which finds its way to the cope

surface of the casting as dirt sand inclusions. It can come from the gating system or in the mold cavity. The causes are too low a hot strength, too dry a molding sand, poor gating design, a deep drop into the mold, washing at the point of impact, metal washing over a sharp edge at gate, metal hitting against a core or vertical wall during the pouring.

36. **Inclusions:** Dirt, slag etc. This defect is caused by failure to maintain a choke when pouring, dirty molding, failure to blow out mold properly prior to closing, sloppy core setting causing edges of the print in the mold to break away and fall into the mold. The drag should be blown out, the cores set and blown out again. Dirt falling down the sprue prior to the mold being poured or knocked in during the weighting and jacketing. For the most part, it's just dirty molding.

Chapter 12
Nonferrous Melting Equipment

Nonferrous metals are melted in numerous types of furnaces. The crucible furnace is the most common, and lowest in initial cost.

The furnace is essentially a refractory-lined cylinder with a refractory cover, equipped with a burner and blower (Fig. 12-1) for the intense combustion of oil or gas. The metal is melted in a crucible (pot) made of clay and graphite, or silica carbide, which is placed in the furnace. When the melting is complete, the furnace is turned off, the furnace cover is opened, and the crucible removed with tongs and placed in a pouring shank. Then the liquid metal is poured into prepared molds.

CRUCIBLE FURNACE CONSTRUCTION

A crucible furnace can be made from an old metal drum or metal garbage can, a few pipe fittings, the blower from an old vacuum cleaner, and thirty or forty dollars worth of castable refractory material. The refractory suppliers can also furnish you with complete advice on what to buy for your particular purpose, along with tips and how to handle the material. Crucibles come in sizes from as small as your thumb to a number 400, which will hold 1200 pounds of molten bronze.

Homemade melting furnaces are simple to construct, and are fun to build and operate. A furnace for a No. 20 crucible, which has a single pour capacity of 60 pounds for bronze, or 20 pounds for aluminum, would be considered average in size.

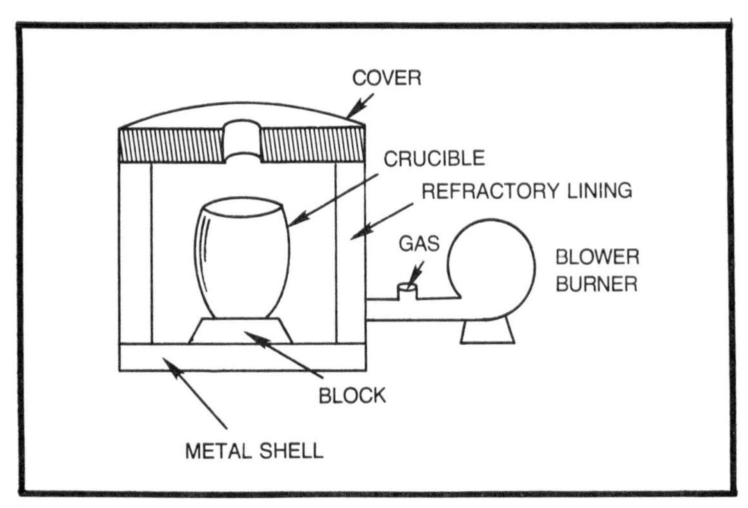

Fig. 12-1. Crucible furnace.

By weight, crucibles will hold three times as much bronze as aluminum. Always allow some clearance below the top of the crucible for safety—it is dangerous to melt a brimming potfull.

Let's assume, for this discussion, that you have chosen a No. 20 crucible: 10⅛ inches high, 7 13/16 inches across the top and bottom, and 8½ inches at its bilge, or widest point (its shape resembles that of a barrel). This crucible requires a base in the form of a truncated cone 6 inches across the top, 7 inches across the bottom, and 5 inches high. Now we have 15 inches of height and 8 inches in diameter to go into the furnace.

There must be sufficient distance between the crucible and the furnace lining for correct combustion, and for room to fit open tongs around the crucible. Also, enough space between the furnace bottom and the covers is needed for correct combustion and exhaust through the cover opening. A good general rule here is to allow 2½ to 3 inches of clearance between the furnace wall and the bilge, and 3 inches between the top of the crucible and the cover and above the furnace bottom. In general, the lining should be a minimum of 4 inches thick to insure good insulation.

With this we need a shell for our furnace 22½ inches (inside diameter) by 22⅛ inches high, with a cover band (Fig. 12-2) 22½ inches (inside diameter) by 4 inches in height. A safety hole directly in the front of the furnace, flush with the bottom and 3 inches square, is needed. Should the crucible break, the metal would run out of the furnace through the hole.

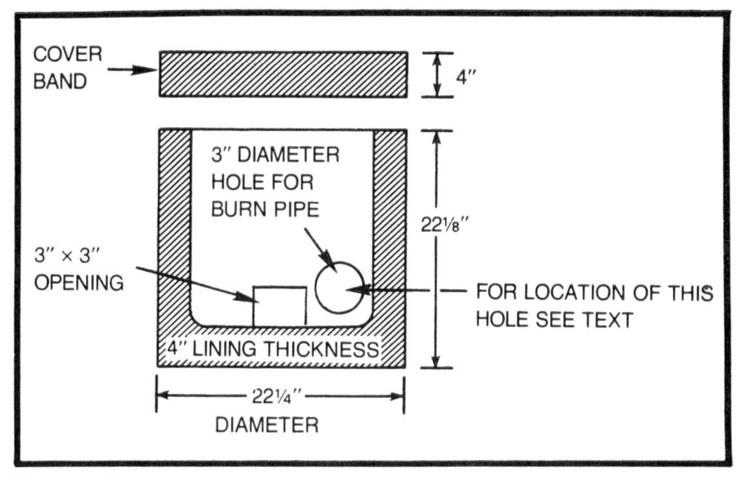

Fig. 12-2. Furnace dimensions.

Without a safety hole, metal could run into the burner pipe, or simply fill up the bottom of the furnace and solidify there, leaving you with a hunk of metal next to impossible to remove short of tearing everything apart.

The shell is completed by making a 3 inch hole 6 inches above the bottom of the shell, a third of the shell's circumference away from the safety hole. This is where the burner pipe enters. The burner pipe is brought in 2 inches above the refractory lining of the bottom, and off-center from the diameter of the shell so that the flame coming will circle the space between the crucible and the lining, spiral around the crucible, and out of the vent hole; this gives the highest and most even heat. (See Fig. 12-3.)

The cover consists of a metal band formed into a ring 22½ inches in diameter, but tapered slightly (Fig. 12-4); tabs or lifting

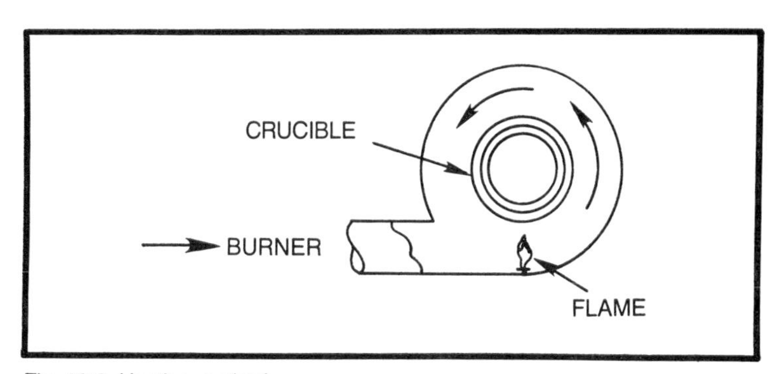

Fig. 12-3. Heating method.

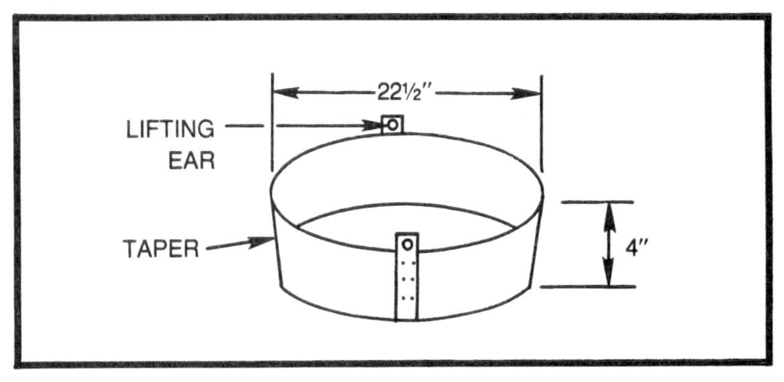

Fig. 12-4. Cover ring.

ears must be riveted on directly across from each other on the cover "ring." They should be drilled with holes to clear a ½ inch pipe, which will be used to remove the cover.

The cover ring is placed on a smooth floor covered with newspaper in preparation for making the exhaust hole.

Within the center of the cover ring, place a 6 inch tin can or jar as a form for the exhaust vent. The castable refractory is now mixed according to the manufacturer's directions, tamped firmly in place around the vent form, and leveled off smoothly with the top of the ring, and left to set overnight.

Now we are ready to line the furnace body. Place a heavy cardboard sleeve 14½ inches in diameter and long enough to extend slightly above the top of the shell in the center of the furnace; 22 or 23 inches tall should do.

Once the sleeve is centered, fill the inside of the sleeve with dry sand to give it added strength, and hold it in place while tamping in the lining all the way to the top. Two plugs will be needed for the furnace shell while the lining is being installed: one to fit through the safety hole and against the cardboard sleeve, and another to fit the burner port; dimensions for the plugs are given in Fig. 12-5.

Coat each plug with heavy oil or grease, and fit both into their respective places. Use small wooden wedges to get a snug fit.

With both plugs securely in place, start tamping in the lining, making sure the castable refractory is well compressed. Do not place too much material between the sleeve and the shell at a time; doing so will produce spaces. When the top of the shell is reached, trowel and smooth the refractory. The furnace is now complete.

Now we must have a suitable burner and blower—a blower that will deliver 300 cubic feet of air per minute.

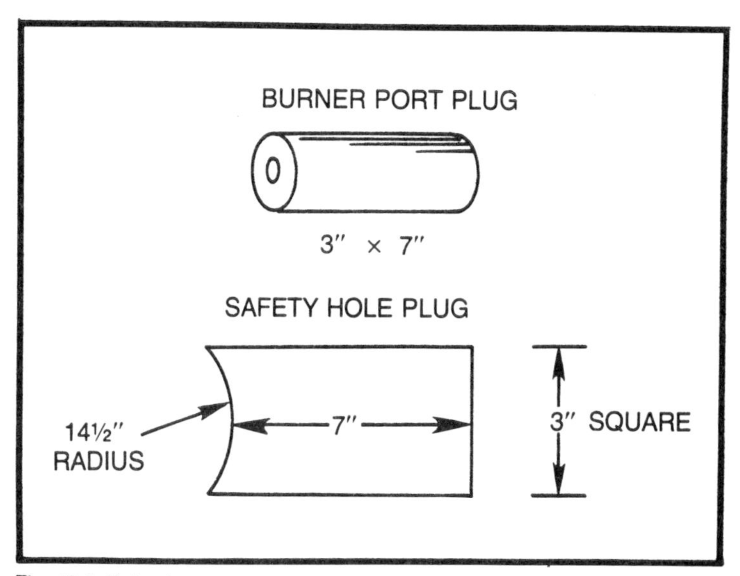

Fig. 12-5. Hole plugs.

A good blower, particularly one from an industrial vacuum cleaner, will deliver enough air at the right pressure.

The simplest type of gas burner can be made from a pipe 2 feet long. Heat the pipe red-hot at one end and hammer the end partially closed—not much over ¼ inch in all around—so that you are left with an opening 1½ inches wide. Heat the pipe again, a foot from the hammered end, and hammer in a neck with an inside diameter 1½ inches that is an inch wide (Fig. 12-6). In the bottom of the groove left from this "necking down," burn a hole and fit a ½ inch pipe snugly into it; weld it in place. You have just made a nipple for the gas line.

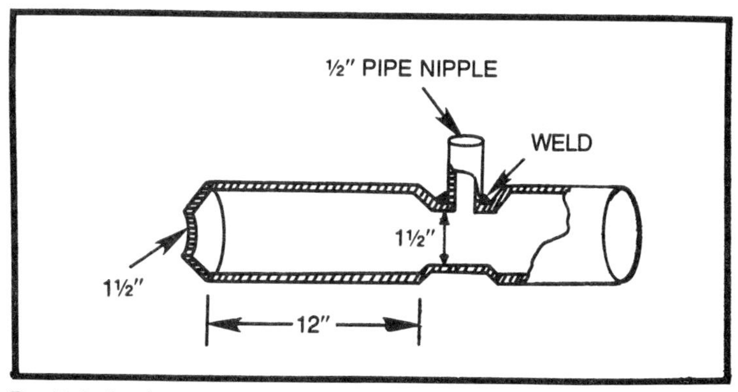

Fig. 12-6. Gas burner.

Insert the burner pipe into the burner port in the furnace body, attach the blower to the opposite end, connect a stopcock to the nipple, and connect the whole thing to a gas outlet.

The intake of the blower must be provided with a shutter or damper to regulate air going to the burner.

A butterfly valve will fit within the burner pipe. Whether you decide to control the air in the burner pipe or at the blower intake doesn't matter; but it must be constructed in a manner that provides positive action without moving of its own accord.

The newly lined furnace should be slowly dried by building a wood fire inside and letting the wood burn down to coals. It takes about two days of this treatment to be safe.

For the first heat, put two thicknesses of cardboard on the crucible support block and place the crucible on the cardboard. With the furnace cover off, place the metal charge loosely in the crucible. Do not wedge the metal in, it must have room to expand without restriction. Wedging metal in a crucible can cause it to split.

Now place a wad of gunnysack material dampened with fuel oil, or charcoal lighting fuel, about a foot from the burner port, in line with the firing direction. The lighter wad should be jammed snugly between the support block and the furnace wall. This will prevent it from being blown out of the furnace or away from the burner. The wad has to remain in place, burning until the furnace wall reaches ignition temperature. Prepare to fire the burner by opening the blower's air control valve halfway. Light the wad and allow it to burn briskly. The blower is started up. Check to see if the wad is still burning briskly. Should it blow out, turn off the blower and start over. Once you have determined that the lighter wad will burn with the blower on, open the gas valve until you get ignition.

Adjust the gas to the point that produces maximum ignition, the loudest roar in the furnace. At this point, you have maximum combustion for the blower's output. Allow the furnace to run for 5 minutes at this setting. After 5 minutes have elapsed, the furnace wall should be hot enough to maintain combustion. Place the cover on the furnace, and advance gas intake and blower output to the point where the gas is wide open and the air is adjusted for maximum roar. Now advance the blower output slightly. This will give you a slight oxidation in the furnace, the best condition for melting.

During the lighting up and the 5 minute period with the cover off, keep your hand on the gas valve. Should you lose ignition during this period, close the gas valve at once, let the blower run a minute or two to clear the gas—air mixture, then close down the blower and start over.

Never relight a furnace using a hot wall; always use a lighter wad.

Never light a furnace without the blower delivering at least half its capacity; too low a setting can result in a backfire into the blower. (This is due to back pressure in the furnace.) The blower has to be blowing strong enough to overcome back pressure so that the ignition takes place in the furnace. Do not light a furnace with the cover on for the same reason. Should the power fail, and the blower stop, immediately turn off the gas. Never leave a furnace unattended during the melting process. To shut down the furnace, close off the gas first, then the blower. Oil-fired furnaces (Fig. 12-7) are lighted and shut down the same way.

COKE FIRED FURNACE

Although there are not many in use today, because they are slow and messy in operation, natural-draft coke-fired furnaces are simple in construction and have a low initial cost.

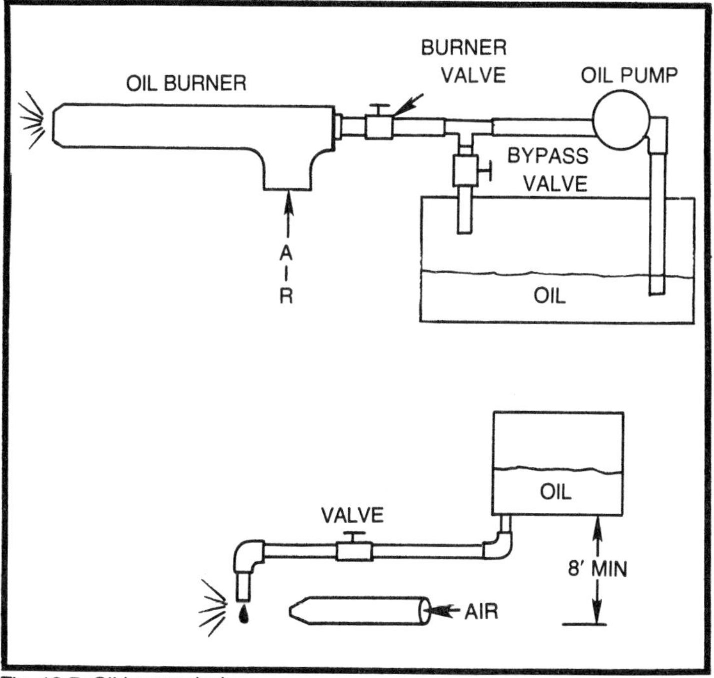

Fig. 12-7. Oil burner designs.

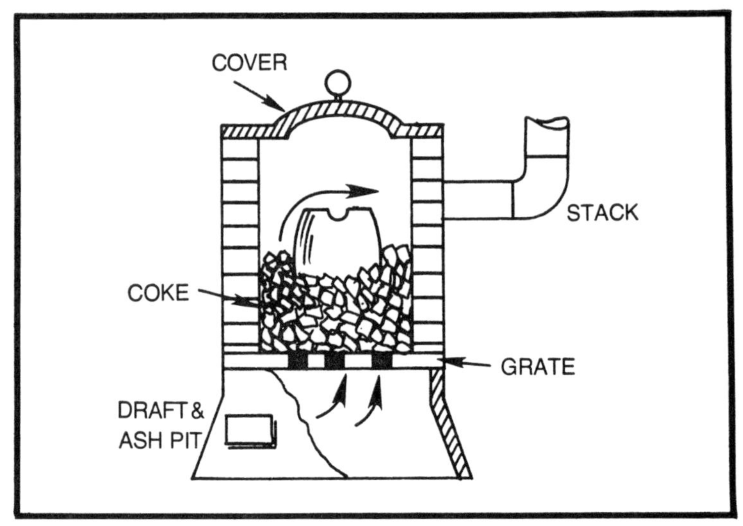

Fig. 12-8. Coke burner.

To operate a coke-fired furnace (Fig. 12-8), place a fire brick on the grate, cover the grate with wood and paper kindling, and cover the kindling with a 3 inch layer of coke. Light the kindling, and when the coke ignites and glows red, add another layer. Add layers until you have a deep bed of red-hot coke. Remove some coke from the center and place the crucible in this well; then build up a layer of fresh coke to the top of the crucible. Fill the crucible with metal and put the cover on. Adjust the draft to promote maximum combustion. As the coke burns down and drops into the ash pit, replace it.

When the metal has reached the desired temperature, pick out coke from around the crucible, and pull out the crucible with the tongs; place it in the shank and pour. The brick in the bottom on the grate is used to support the crucible should it work its way down too far in the furnace.

GROUND FURNACE

A crucible furnace can be easily constructed in the ground (Fig. 12-9). In areas where a water table is not too close to the surface.

TAPPED CRUCIBLE

The tap-type crucible furnace is commonly used by people who work alone, casting small to medium-sized work, or who have a minimum of headroom to pull out the crucible. The furnace (Fig. 12-10) is built on four legs with a front opening at the bottom. This

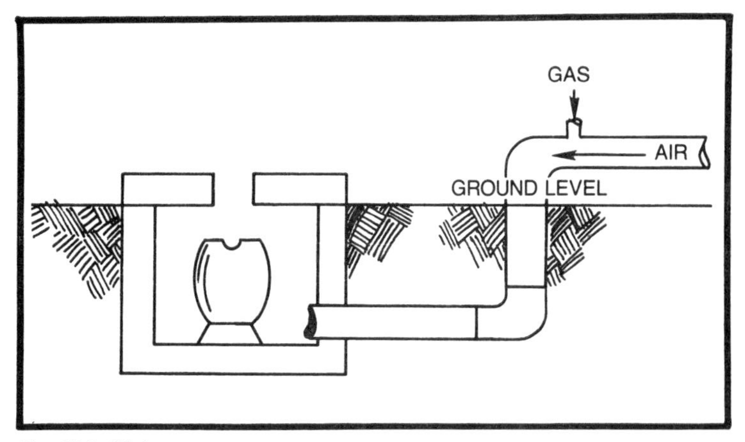

Fig. 12-9. Pit furnace.

opening should be at least 4 inches by 4 inches. A hole is drilled in the front of the crucible, level with the bottom, to allow the metal to run out.

The furnace is lit and charged in the conventional manner. After 5 minutes the brick is removed, the cover is closed, and combustion is adjusted. A refractory-lined ladle is preheated by placing it on bricks above the exhaust hole in the cover during the melting.

When the metal is ready to pour, the hot ladle is removed with a hand shank (handle) and placed in front of the furnace.

A clay plug in the crucible's tap hole is picked out with a tapping bar made from a ½ inch steel bar sharpened to a point; the metal runs into the hand ladle. The ladle is then used to pour the mold.

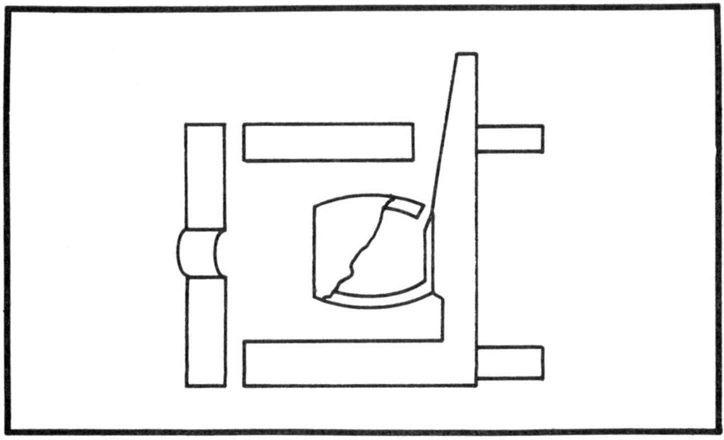
Fig. 12-10. Tap type furnace.

To close up the tap hole for the next heating, a cone of bod (plug) mix is formed on a bod rod, and firmly pushed into the hole. The bod rod is given a little twist to release the bod. The heat of the furnace and crucible bakes the bod into place.

The crucible is placed in the furnace with the tap hole facing the opening in the front of the furnace. A refractory trough is made from the bottom of the tap hole to the furnace opening to carry the tapped metal out to the ladle. The operation is very simple. The tap hole is stopped with a bod made of 1 part sharp silica sand to 1 part milled fireclay; the mixture is dampened to form stiff mud and pressed firmly into the hole. To light the furnace, the lid is simply propped up at its edge with a chunk of firebrick. If the hole should weep or leak at any time during heating (very rare), put a ball of bod mix on the bod rod and press it over the leak.

POURING THE MOLD

The actual pouring of molten metal into molds is a very important phase of the casting operation. More castings are lost due to faulty pouring than to any other single cause. Some basic rules for "gravity casting" a mold poured from a crucible or ladle are:

1. Pour with the lip of the ladle or crucible as close to the pouring basin as possible.
2. Keep the pouring basin full (choked) during the entire pour.
3. Keep the pouring lip clean to avoid dirt or a double stream.
4. Use slightly more metal than you think you'll need.
5. Pour on the hot side—more castings are lost by pouring too cold, rather than too hot.
6. Once a choke is started, do not reduce the stream of metal.
7. Do not dribble metal into the mold or interrupt the stream of metal.
8. If a mold cracks and the metal starts to run out, don't try to save it.
9. If a mold starts to spit metal from the pouring basin or vents, stop pouring. Continuing to fill a wet mold that is spitting back can result in a bomb.
10. Don't use weak or faulty tongs, or shanks.
11. Keep the pouring area clean and allow plenty of room for sure footing and maneuvering.
12. Do not pour with thin, weak crucibles.
13. Wear a face shield and leggings.

14. When pouring at night, dust the pouring basin with wheat flour in a bag to make a more visible target for the pouring.
15. When pouring with a two-man shank, make sure the other man knows what he is doing (not move or jerk back once you have started the pour).
16. Make sure you have a good, dry pig bed to hold any excess metal after the pour.
17. When pouring with a hand shank, rest the shank on your knee.
18. When pouring several molds in a row with a hand shank, start at one end and back up as you go. Going forward to pour brings the knuckles of the hand closest to the ladle over the mold just poured.
19. When pouring several molds from a single ladle or crucible, pour light, thin castings first (the metal is getting colder by the minute).
20. Don't try to pour too many molds at a single heating.
21. Make sure that flasks are closed and clamped or weighted properly.
22. Don't pour when you are in an awkward position. You must be relaxed.
23. Don't try to lift and pour too much metal by hand. Use a crane or chain fall on a jib boom. More than 40 pounds in a hand shank, or 200 pounds in a two-man bull shank begs disaster.
24. If the metal in the ladle or crucible is not bright, clean, clear, and hot, don't pour it.

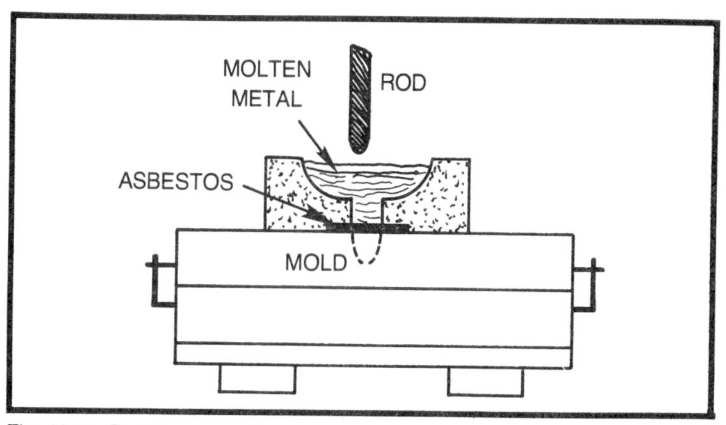

Fig. 12-11. Pouring reservoir.

Large molds are often poured by placing a sheet of asbestos over the sprue, on top of which is a large molten metal reservoir made of sand. The amount of metal required for the casting is poured into the reservoir, and a rod is used to puncture the asbestos sheet (Fig. 12-11), allowing the metal to fill the mold.

Chapter 13
Nonferrous Casting Metals

The nonferrous metals predominately used for casting are alloys of copper; that is, brass and bronze. Brass is generally identified by its color, and is said to be either red or yellow; color can serve as an indication of the temperature required for the melt.

FLUXES

When a copper-based metal is melted in a crucible, some of the metal will combine with oxygen to form cuprous oxide. To convert the cuprous oxide back to metallic copper, something has to be added to the melt to draw the oxygen. The one most commonly used for red metals is phosphor-copper: an alloy of copper and phosphorous. Phosphorous has such an affinity for oxygen that it will ignite upon exposure to air; to make it stable enough to use, it is alloyed with copper in the form of "shot." When introduced into the molten metal, the copper melts and releases the phosphorous, which deoxidizes the cuprous oxide in the melt.

Oxygen can be carried out of a melt by nitrogen. The nitrogen gas (dry) is introduced through a hollow tube called a lance, which is connected to the gas cylinder by a rubber hose. For red metals, the lance is carbon; for aluminum, it is iron or steel. The tube is inserted into the molten metal to within an inch or so of the bottom of the crucible.

You can "roll your own" fluxes, so to speak, in several ways. A very good deoxidizer for copper is 5 ounces of black calcium boride powder sealed in a copper tube 2 inches long, with the ends crimped closed. The best flux for melting down extremely fine scrap pieces, such as buffings or grindings, is plaster of Paris.

There is another group of fluxes that are used to prevent the products of combustion in the furnace from coming into contact with the molten metal and oxidizing it. Because they are used as a protective cover, they are called cover fluxes. Such a flux can be made by mixing 5 parts ground marble with 3 parts sharp sand, 1 part borax, 1 part salt and 10 parts charcoal. Or, try this one: Mix equal parts of charcoal and zinc oxide, add enough molasses water to form a thick paste; roll the paste into 2 inch balls and let them dry. When the alloy starts to melt, drop in enough balls to cover the surface.

Some fluxes are very detrimental to refractory linings, often eating away a ring around the inside of a crucible during a single heating. Check with your crucible and flux supplier to find out what is compatible before using a particular flux.

RED BRASS

Leaded red brass can be used with a simple gating system but it must be choked because these alloys flow quite freely. Risers are required for heavy casting sections, and the melt must be fast (under oxidizing temperatures). No cover flux is necessary for these alloys, especially when clean materials are used; deoxidize with 1 ounce of, 15 percent phosphor-copper, for each 100 pounds of melt. Too much deoxidizer will make the metal too fluid, and can result in dirty castings.

The average composition of red brass is: 85 percent copper, 5 percent tin, 5 percent zinc, and 5 percent lead; this alloy is commonly called eighty five and three fives or "ounce metal." Semired brass is handled like red brass. Its composition is normally; 78 percent copper, 2.5 percent tin, 6 percent lead, and 7 percent zinc. The pouring temperature range for this alloy is 1950 degrees Fahrenheit for very heavy sections, to 2250 degrees Fahrenheit for very thin sections. Generally, 2150 degrees Fahrenheit can be considered as an average pouring temperature.

Using leaded yellow brass necessitates gating similar to that used for red brass, with the exception that the sprue, gates, and runners must be somewhat larger. The mold cavity must be filled as rapidly as possible. If filled too slowly, the zinc in the alloy will produce a "wormy" surface on the casting. If melted in an open-flame furnace, such as a rotary or reverberatory type, high zinc loss will result. Crucible melting is best for yellow brass. The general pouring temperature is 2050 degrees Fahrenheit. No cover flux is required. But, the metal should be deoxidized with 2 ounces of aluminum per 100 pounds of melt. (Never use aluminum and phosphor-copper together).

To prevent zinc from condensing in the mold cavity, the cause of the surface condition described, tip the mold so the sprue is at the low end of the mold (Fig. 13-1).

The normal composition of yellow brass is 74 percent copper, 2 percent tin, 3.5 percent lead, and 20.5 percent zinc. "High-strength" leaded yellow brass (manganese-bronze) is characterized by high shrinkage, and the tendency to form dross (oxides) during pouring, or when agitated. Bottom horn gating is preferred, and large risers and chills must be used. It's a tough metal to cast, and requires considerable experience. It is best melted in crucibles and poured at the highest possible temperature to prevent the excessive production of zinc fumes. High temperatures are also recommended to reduce the risk of flaring; flames shooting up from the surface of the molten metal. Yellow brass will flare at about 1850° Fahrenheit. Approximately 1½ pounds of zinc will be lost for every 100 pounds of alloy melted; this must be replaced. The composition of the high-strength alloy is 62 percent copper, 1.5 percent man-

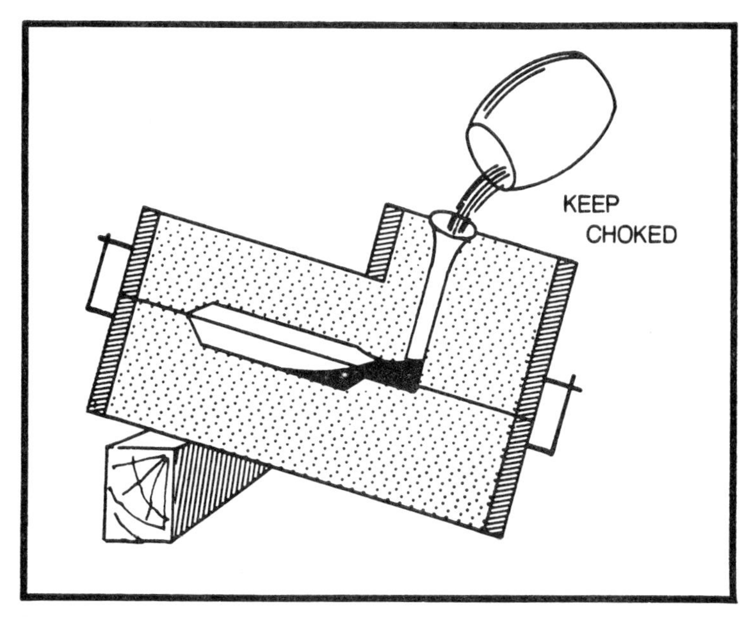

Fig. 13-1. Pouring yellow brass.

ganese, 1.5 percent aluminum, 2 percent iron, 1.5 percent tin, 0.5 percent lead, and 31 percent zinc.

To test for the presence of aluminum in brass, dissolve 5 grams of metal in a solution of 1 part nitric acid to 1 part distilled water. If there is a white precipitation (tin), filter it out. Then, neutralize the solution with ammonium chloride, until it turns blue, and filter it. If aluminum is present, there will be a gelatinous precipitate on the filter paper. To test for the presence of silica in brass, dissolve 5 grams of metal in the same solution and let it boil for about 10 minutes. If silica is present, it will show up as a gelatinous precipitate around the edge of the container.

BRONZE

Tin-bronze and leaded tin-bronze are very widely used for art and architectural castings, especially "88-10-2", known as straight bronze (88% copper, 10% tin, 2% zinc). It is also known as navy G or M bronze.

These alloys require risers on heavy casting sections because of their shrinkage characteristics. On light, uniform castings, a choke gate (Fig. 13-2A) is recommended; a skimmer gate (Fig. 13-2B) will be needed for a heavy section. The melt should be rapid (slightly oxidizing). Raise the temperature only slightly above the temperature required to pour the mold. Fluxes are not usually necessary, but if desired, a charcoal cover layer will do. Deoxidize with 2 ounces of 15 percent phosphor-copper per 100 pounds of melt. The pouring range is 1950 degrees Fahrenheit for heavy castings, to 2250 degrees Fahrenheit for very thin ones; 2150 degrees Fahrenheit is average. Straight bronze takes a fine patina and can be easily chased and finished.

Silicon-bronze has come into favor for art castings during the past decade. There are eleven different silicon-bronze alloys available, the one most used being 94.5 percent copper, 4 percent silica, and 1.5 percent iron. Silicon-bronzes had various trade names when they were under patent. The names still appear as the trademarks P.M.G., Tombasil, Everdur, and Herculoy. All four have very similar characteristics.

Silicon-bronze is compatible with any lead-free brass or bronze, such as 88—10—2 bronze. When you melt several alloys in your operation, avoid contamination between silicon-bronze and any brass or bronze bearing lead. Use separate crucibles and keep all gates and returns separate.

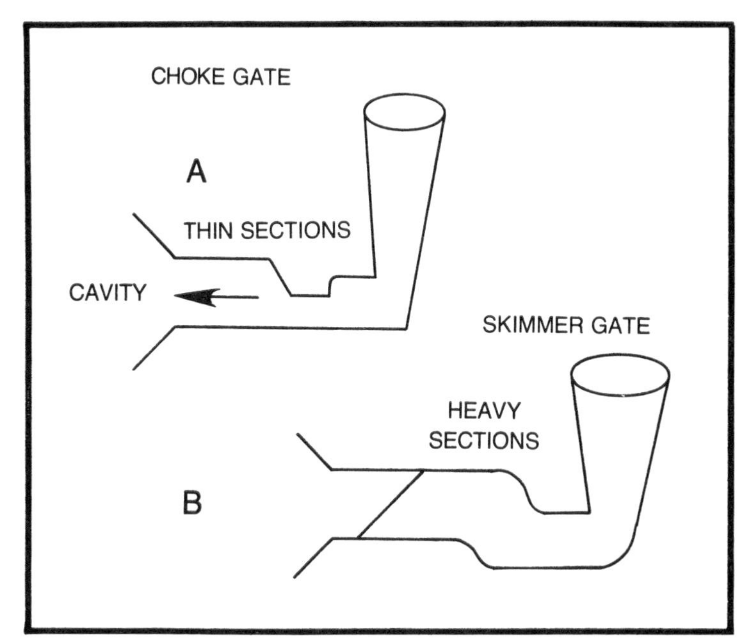

Fig. 13-2. Gating for bronze.

Silicon-bronze can be easily welded with oxyacetylene or electric arc methods. Use silicon-bronze rods or flux-coated, lead-free yellow brazing rods. Silicon-bronze can also be easily welded to steel.

MAKING COPPER ALLOYS

When making alloys from scratch, the order of melting should always be: melt the copper, add any clippings or punchings (punched-out scraps), then add zinc, lead and tin. A good mix for general casting is 71½ pounds of sheet brass clippings, 25 pounds of copper wire, 2 pounds of copper wire, 2 pounds of lead, 1 pound of tin, and 8 ounces of 15 percent phosphor-copper. For yellow brass mix 60 pounds of copper wire or sheet copper with 36 pounds of zinc and 4 pounds of lead.

Red brass can be formulated cheaply from 40 pounds of copper wire, 7½ pounds of zinc, 7½ pounds of lead and 45 pounds of brass scrap (valves, etc.); an alternative to this is 83 pounds of copper wire, 8½ pounds of lead, 6½ pounds of zinc, and 2 pounds of tin.

A good, tough alloy that will not crack when bent double can be had with a mixture of 84 pounds of copper wire (or clean, copper scrap), 10½ pounds of zinc, 3 pounds of lead and 2½ pounds of tin.

There are over 40 bronzes that can be called statuary bronzes; some are more on the order of yellow brass. The alloy coming closest to being considered "standard" for statuary work is 90 percent copper, 7 percent tin, and 3 percent zinc, by weight. The standard alloy for grave markers and reliefs is 88 percent copper, 6 percent tin, 3.5 percent zinc, and 2.5 percent lead. The government prefers its own statuary and plaque bronze: 90 percent copper, 5 percent tin, and 5 percent zinc. (This is approximate; 8 ounces of lead are added to every 100 pounds of the molten alloy.)

It has been found through analysis, that the composition of alloys used in famous statues throughout the world was influenced by the individual preference of the sculptor.

Various kinds of bells call for specific alloys, some compounded for tone. Railroad signal bells get their peculiar clang, in part, from a mixture of 60 percent copper, 36 percent zinc, and 4 percent iron. The alloy used for gongs contains no zinc: 82 percent copper, 18 percent tin. Likewise, very large bells are generally made from just copper (76%) and tin (24%), as are sleigh bells, in a ratio of 40 to 60. Fire engine bells are 20 percent tin, 2 percent nickel, 1 percent silicon, and 77 percent copper. Many would concur that a bell 78 percent copper and 22 percent tin has the best tone.

Silicon-bronze makes a strong bell with an excellent tone that can be heard for quite a distance.

ALUMINUM ALLOYS

Aluminum alloys are melted and handled much the same as copper-based alloys; melt under oxidizing conditions, and gate for progressive solidification with a minimum of turbulence. They have, however, a fairly high rate of shrinkage during solidification; attention must be paid to correct risering to prevent this. It is common to increase the strength of an aluminum casting by as much as 50 to 100 percent by redesigning or relocating the gates and risers. Cores must be low in dry strength and high in permeability. The pouring temperature range is usually between 1250 to 1500 degrees Fahrenheit. Deoxidizing is done with solid fluxes, or by bubbling nitrogen through the molten metal. Aluminum is melted in crucibles, cast iron pots, and in open flame furnaces.

Casting With Scrap Metals

Making castings from scrap metals is frowned upon in many camps but the producers of ingot metal do just that. Ingot metal is preferable to use because a chemical analysis is furnished with each shipment. It is clean and easy to handle. However if you have a good source of clean, graded scrap available, take advantage of it. If a casting does not require an exact "pedigree" or stringent strength characteristics, and is nonfunctional, why not use good scrap?

Certain precautions should be taken when scrap metal is melted. It should be free from oil, dirt, paint, iron, steel, aluminum, and other impurities. Scrap valves should be carefully inspected, and any silicon-bronze or aluminum-bronze seats removed. Using extremely thin scrap will result in high metal loss due to rapid oxidation. Melt heavy scrap first to form a "heel" of metal under which lighter metal can be pushed in small amounts.

Let's assume you're using a crucible melt of scrap consisting of copper wire, sheet brass clippings (or punchings), lead and tin. First, melt the copper wire under a cover of charcoal with the furnace adjusted to a slightly oxidizing condition. When the copper melts, add 6 ounces of 15 percent phosphor-copper to each 100 pounds of melt. Now place a ring on top of the crucible made from an old crucible (Fig. 13-3), so that a large quantity of material can be packed in each time. When this material becomes red-hot, push it under the molten copper.

Repeat this process until the complete charge has been melted. Then, add tin followed by lead. Bring the metal to 100 degrees Fahrenheit above the desired pouring temperature, deoxidize with 2 ounces of phosphor-copper shot for each 100 pounds of melt, skim, and pour.

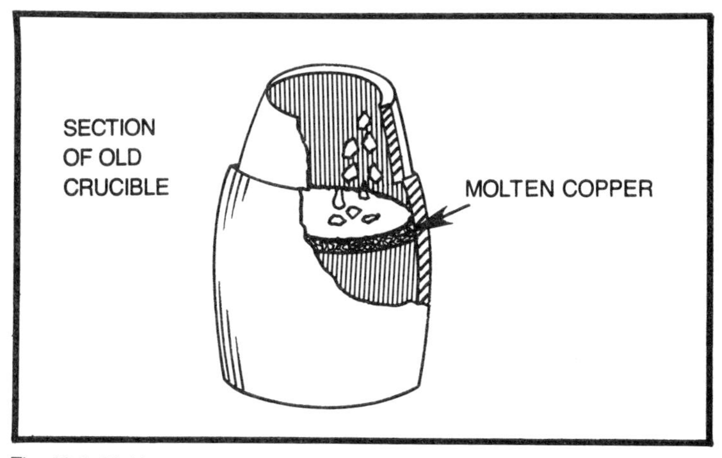

Fig. 13-3. Melting scrap.

To derive the metal you desire, approximate its analysis, and use the weight percentage brass scrap analysis. With a wide enough selection of scrap, just about any metal alloy you desire can be compounded.

The selection of scrap must be made more carefully by the casting maker than by the ingot maker. The ingot maker melts in large quantities and can reduce the impurities to acceptable amounts through dilution alone. A five pound chunk of lead in a 20 or 30 ton melt of silicon-bronze is so imperceptible as to be negligible. A quarter of a pound of lead in 100 pounds of molten silicon-bronze will ruin the melt.

Good melting equipment operated correctly, good deoxidizing practices, and an accurate pyrometer are essential to a successful melt. It is impossible to gauge the temperature of molten metal visually within 200 degrees Fahrenheit. Some think they can, but risk a lost casting for the lack of a relatively inexpensive piece of equipment. To sum up the melting of nonferrous metals:

- Melt as fast as possible.
- Melt under slightly oxidizing conditions.
- Don't soak the metal (hold in the furnace) too long.
- Use separate crucibles for each type of alloy to prevent contamination.
- Don't heat over 100°F above the desired pouring temperature.
- Use a good pyrometer.
- When melting scrap be selective: melt only clean scrap.
- Keep melting equipment in good condition and repair.
- Establish a good deoxidizing method and stick with it.
- Handle molten metal very carefully.

Chapter 14
Cast Iron

Countless examples of beautifully made, cast-iron works can be found worldwide. Executed purely in the name of art or to serve a useful function, they range in weight from a few ounces to several tons. One only has to roam the aisles of any antique shop to discover fine representations of this work: cast-iron toys, penny banks that take your coins with the tip of a hat, and of course, friendly, ornamented wood and coal stoves.

Stove plate molders, extremely talented artisans, produced work often unfathomable in the techniques employed. Some castings were not only highly ornamented, but were extremely thin for their total area. The pattern maker who produced the original master not only had to have a knowledge of the foundry, but had to be a master woodcarver as well.

The master patterns were mahogany, highly carved on the face and carefully cut away inside (backed out) to produce a thin, even wall. Carving the face was a delicate enough task, but carving out the back required time-consuming patience that might only be rewarded with a week's work ruined.

Several sculptors today are doing fine work in cast iron, an art drawing a growing interest. You can produce small heats (quantities) of cast iron in your crucible furnace with little trouble. Melting cast iron in a crucible is actually so simple, it's a wonder more people don't do it.

The usual practice incorporates clay and graphite crucibles. The iron, free from contaminants (anything other than iron), is broken into small pieces, the size of walnuts or a little larger. The crucible is charged (filled) with alternate layers of charcoal and iron, to which has been added about two handfuls of soda ash. The soda ash can be placed either below, or on top of the charge. After you have gone this far with the process, cover the crucible with an old crucible bottom and place a circular piece of corrugated cardboard between the base and support block to prevent the crucible from sticking. Start the furnace with the flame set to slightly oxidizing. Gray iron melts at about 2327 degrees Fahrenheit. Although this figure may look large compared to that for the melting point of brass, you will find that the layers of charcoal will promote rapid melting that can be done in not much more than the time it takes for brass. With a little practice you can produce high-grade iron castings. (Old cast-iron steam radiators can be easily broken up to produce an excellent source of very fluid iron.)

THE CUPOLA

The largest percentage of cast iron produced is melted in the cupola (Fig. 14-1), basically a miniature blast furnace. The cupola shown is the basic furnace used to reduce copper ore to matte copper. The cupola, besides reducing ore, can melt bronze, brass, or cast iron. Although the cupola is usually thought of as a cast-iron melting furnace, it is excellent for melting bronze, in particular, silicon-bronze and bronze low in lead or zinc content. I have used a small, homemade cupola for many years for all my melting.

Because the cupola melts continuously—as long as it is stoked and fed—it will melt charge after charge. The big advantage of the cupola is that you can pour a large quantity of molds, and collect a sufficient number of charges in a big ladle to pour castings in just about any weight you wish. You can also melt single batches of cast iron or bronze. Commercial cupolas range in size from a No. 0 with an inside diameter of 18 inches that will melt a ton of iron in an hour, to a No. 12, 84 inches in diameter that will melt more than 33 tons an hour.

A cupola is basically a refractory lined cylinder with openings at the top for the escape of gases and the introduction of a charge. Smaller openings at the bottom are provided for the air blast and the release of molten metal and slag. A bed of fuel is laid in the cupola and ignited, after which alternate layers of metal and fuel are put down

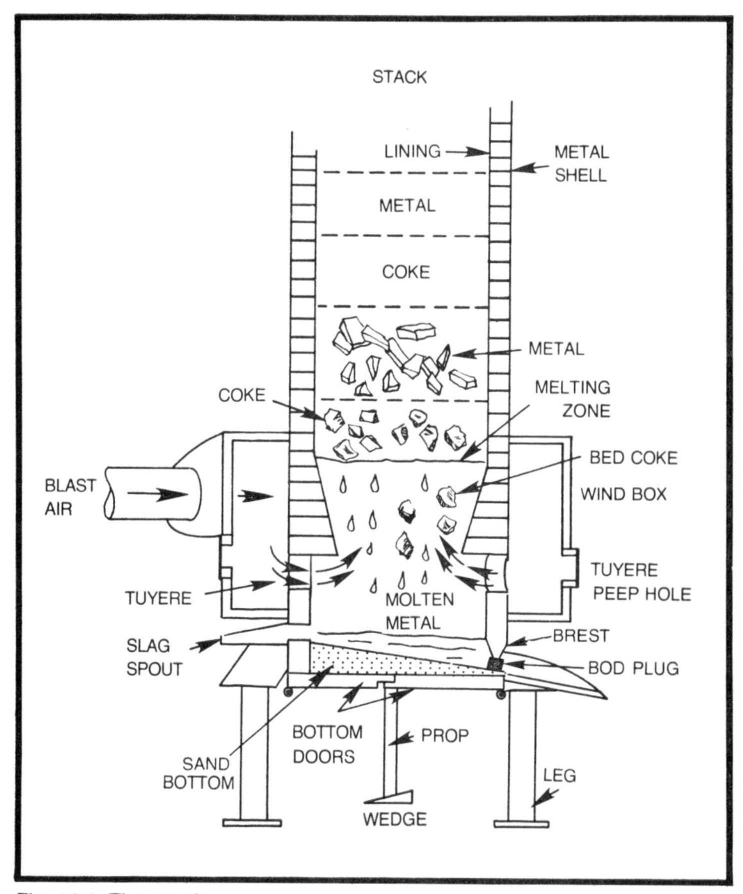

Fig. 14-1. The cupola.

and the blast turned on. If a few simple rules are followed, melting will begin quickly and continue for a long time.

The cupola is by far the simplest and most efficient melting furnace. Before getting into its actual operation,let's examine the cupola's construction from the bottom up.

CONSTRUCTION

The cupola is supported on four legs and fitted with a pair of perforated, cast-iron doors underneath. The legs are long enough to allow the doors 6 inches of clearance above the floor when they are open (hanging down). When the doors are closed, they are held by a metal pipe wedged in place tightly with a pair of steel or cast-iron wedges (Fig. 14-2A). The doors have a mating offset lip (Fig.

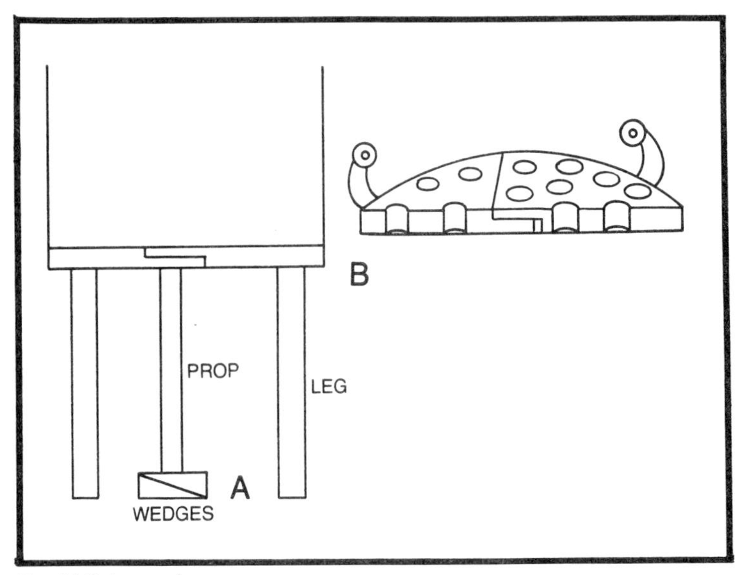

Fig. 14-2. Lower doors.

14-2B) and ledge to insure a good fit. Four or more openings are cut through the lining and shell about a foot from the doors for tuyeres. They fan out at the inside of the lining and make an almost continuous opening by being close to other adjoining tuyeres (Fig. 14-3). The tuyeres are usually formed in cast-iron tuyere boxes installed during the lining of the cupola. To provide a blast of air through the tuyeres, the cupola is constructed with a wind belt—a sheet metal tube encircling the outside of the cupola. There are two basic types. In the most common, air is introduced into the tuyeres through elbow connections extending from the bottom of the wind belt and terminating at the outer tuyere opening (Fig. 14-4A): these are called pendulum tuyeres. In the second type, the wind belt encircles the tuyeres at their level (Fig. 14-4B).

It is customary to place one tuyere lower than the rest; this is a safety tuyere. Its purpose is to prevent molten metal from rising above the bottom of the tuyeres and spilling, burning up the wind belt or causing greater havoc. Should the metal rise too high it will spill over into the safety tuyere which is fitted with a lead disc covering a hole in the bottom of the pipe that supplies the blast; the disc would melt quickly, allowing the metal to drop through, thus preventing damage to the cupola.

A ladle is usually placed on the ground below the disc. Each tuyere has a peep hole fitted with a cover and mica window. The

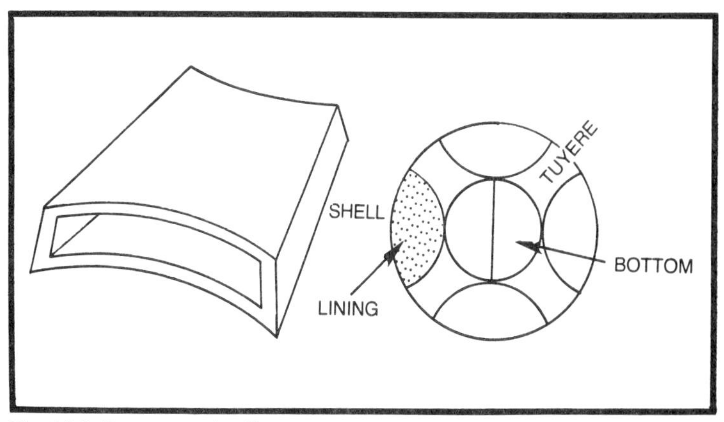

Fig. 14-3. Tuyere construction.

cover (Fig. 14-5) can be swung aside for access to the tuyeres, or locked in place when the cupola is under blast.

Blast air is supplied to the wind belt by a blast pipe. It enters the wind belt at a point between tuyeres. The blast circulates through the wind belt (Fig. 14-6) and enters the cupola through the tuyeres. This type of construction seems to be the best in that the blast is divided more or less evenly, delivering the same volume and pressure to all tuyeres equally. The blast is supplied to the cupola by a blower whose required output is based upon the inside dimensions of the cupola.

Besides tuyeres, the cupola must have a tap hole (Fig. 14-7) through which molten metal can be removed. This is located at the

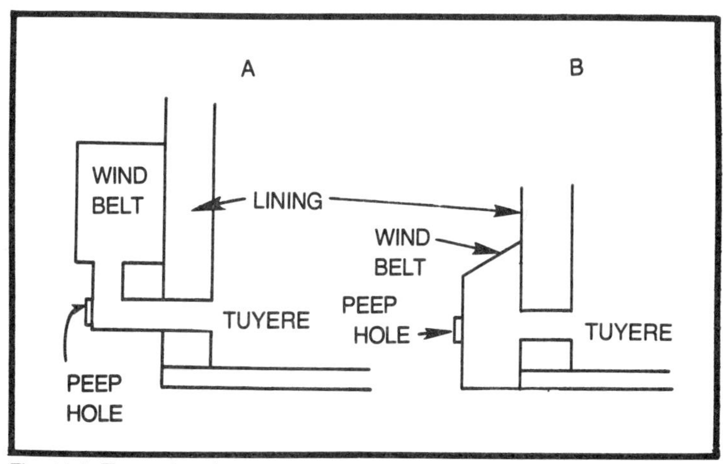

Fig. 14-4. Tuyere locations.

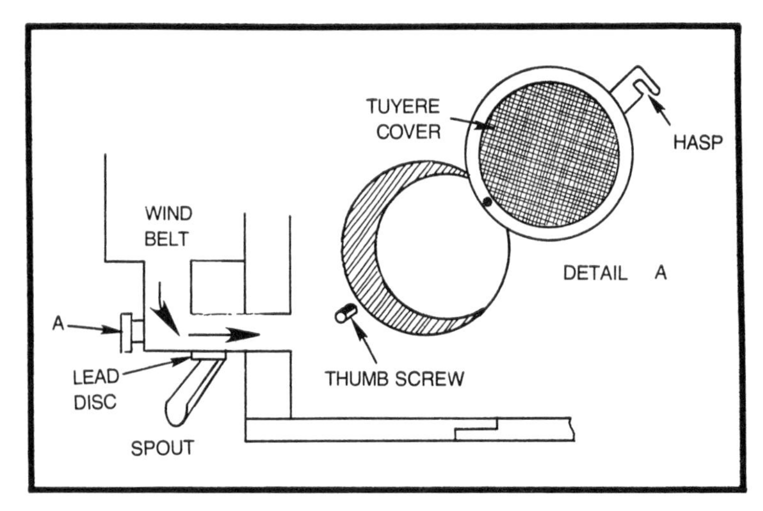

Fig. 14-5. Peep hole cover.

bottom front, between two tuyeres; a slag hole is located in the back between two tuyeres, or 4 inches below them. The legs are sections of pipe attached to a plate used to anchor the cupola to the floor.

A truly workable cupola, and one that will operate as safely and efficiently as possible, must be constructed according to certain design criteria. The refractory lining, for example, should never be less than 4½ inches thick. The diameter of the lining itself is determined by its thickness; if thinner than 7 inches, the inside diameter should be no more than 36 inches—a thickness greater than that permits a diameter from 41 to 56 inches.

Blower output should be 2.5 CFM (cubic feet per minute) per square inch of inside lining area. Always round off the output figure

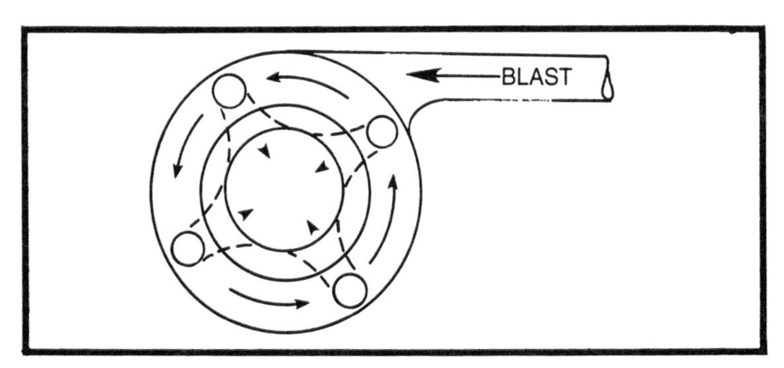

Fig. 14-6. Wind belt.

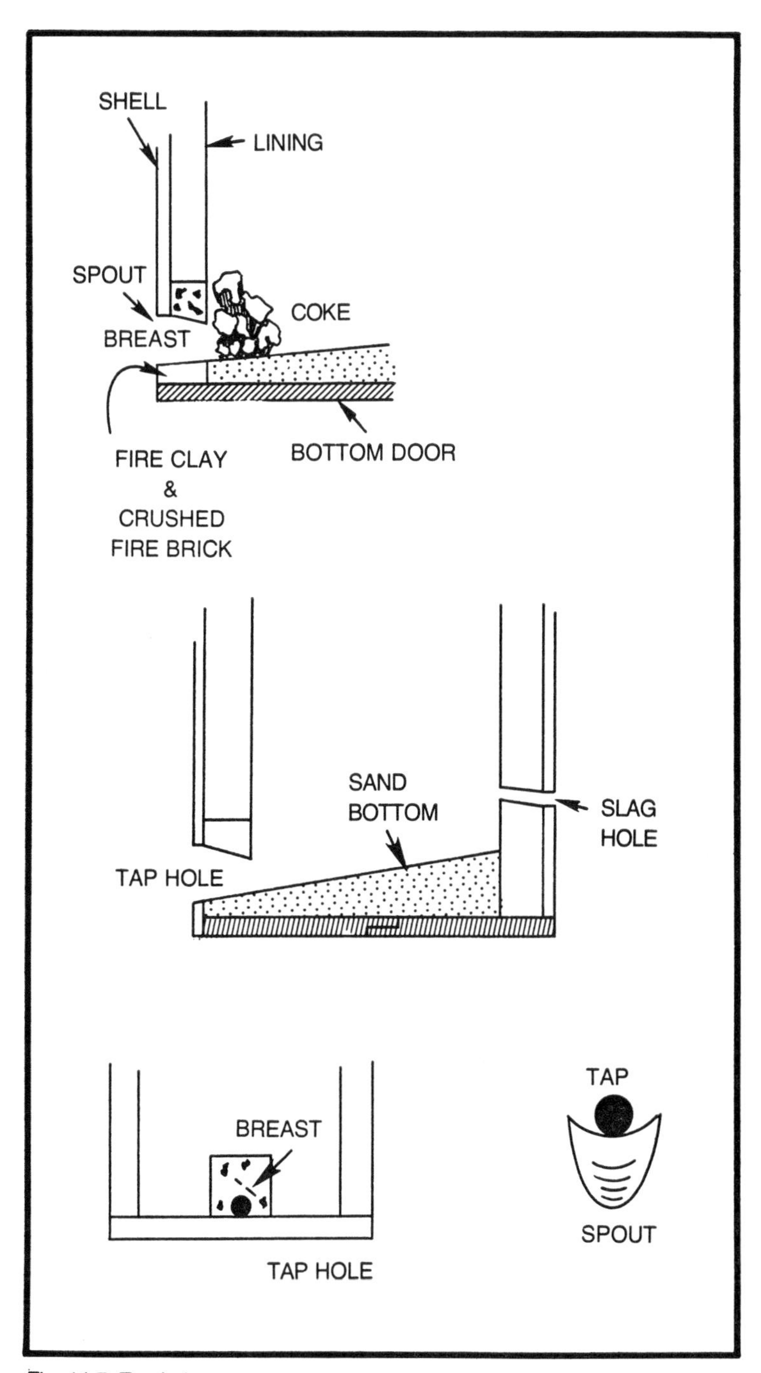

Fig. 14-7. Tap hole.

you derive to the next highest blower size available, for safety. The level of the coke bed above the tuyeres is determined by blast pressure. Its height should be equal to the square root of the blast pressure times 10.5 plus 6.

A good way to determine the correct bed height is to note the length of time it takes the first drops or iron to fall past the tuyeres after turning on the blast. Iron should be seen at the tuyeres in 4 or 5 minutes; at the tap, in 6 to 8 minutes. If iron can be seen sooner, the bed of coke is too low; seen later, the bed is too high. This observation can be made at a tuyere window.

The optimum melting rate is based on 10 pounds of iron per square inch of cupola area; that is, a 12 inch cupola should melt 1130 pounds per hour. The amount of iron needed per total charge—fuel plus metal—is based on a melt ratio, a value determined by the weight of the coke charges. Enough metal melted by the time 4 inches of the fuel bed have been burned away is optimum. A layer of fuel previously placed above the metal charge will replace the fuel consumed. When another charge of metal is melted away, more fuel will descend; the process is repeated until the end of the heat, or until the desired amount of metal is melted.

The constant relating to the weight of metal charges is the weight of 4 inches of coke. Usually, a metal ring 4 inches deep is made that has the same diameter inside as the cupola; it is filled with coke, and then the coke is weighed.

The melt ratio for iron is 9 to 1, nine pounds of iron for each pound of coke. If a cupola took 5 pounds of coke to reach a 4 inch level, each iron charge would weigh 45 pounds (each coke charge would be 4 inches high. The melt ratio for copper or bronze is 20 to 1.

Some small cupolas, 24 inches tall and 24 inches wide, are used for batch melting (one charge of metal).

These extremely short cupolas have a cover similar to those on furnaces used to melt bronze, and have their tuyeres close to the bottom. After each melt, the bed is built up and a new metal charge is added. This is done throughout the day, whenever a melt is needed. The cover has a 6 inch exhaust hole in its center. The cover provides back pressure to drive the blast downward to compensate for the cupola's height.

No two cupolas operate alike, but the guidelines given will put you in the ballpark; the rest of the pudding is in the eating. The most common type of homemade cupola is two 55-gallon oil drums welded end to end. They will do a remarkable job if built well and properly operated.

OPERATION

After raising the bottom doors and propping them up firmly, tamp down a compact layer of tempered molding sand on the bottom of the cupola, making it slope downward from the back and sides towards the tap hole. Cover the bottom with light, dry kindling and light it.

Add enough wood to get a good brisk fire going. Add approximately half of the bed coke; when this has burned brightly (as seen through the top), add enough coke to bring it to the correct level. At this point, put in the first metal charge, followed by 4 inches of coke. Then add the next metal charge and so forth, until the desired number of charges (Fig. 14-8) are in the stack. Allow the cupola to "soak" for about a half hour. (Whenever the blast is off, open one or more tuyeres to prevent the accumulation of explosive gases in the wind belt).

The bed should now be properly burned through and the cupola charged with alternate layers of metal and coke. Close the tap hole with a bod (clay plug). The bod mix is fire clay and sand in equal amounts, tempered with enough water to make a pliable mixture. Make the bod on the end of a bod rod (Fig. 14-9) by first wetting the

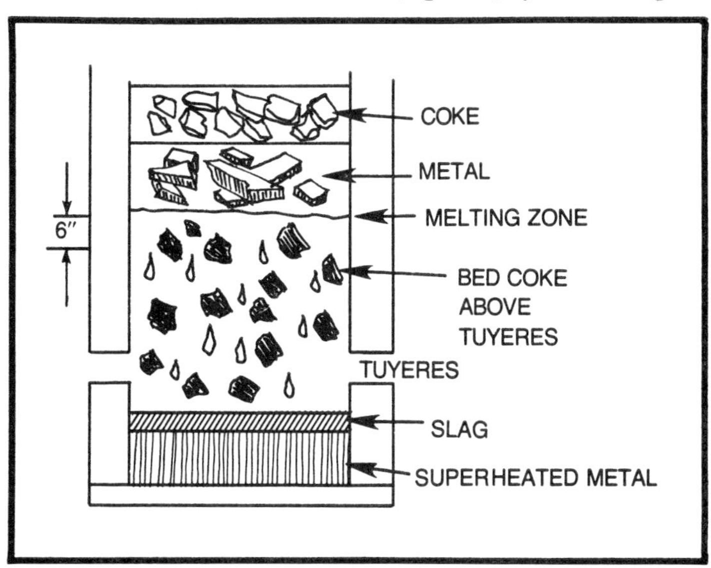

Fig. 14-8. Charges in stack.

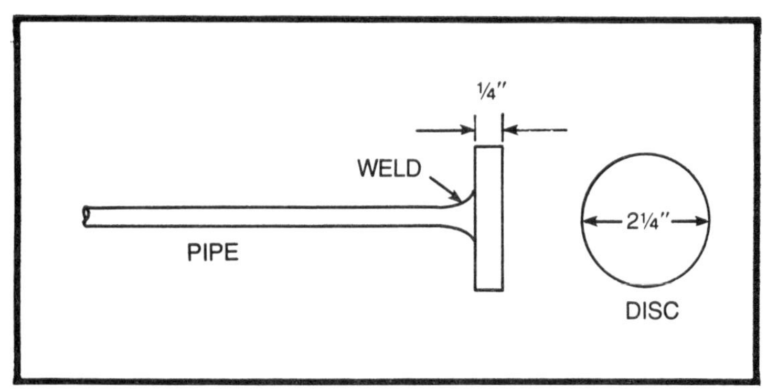

Fig. 14-9. Bod rod.

pad on the bod rod and forming a ball of bod mix there; then form it into a cone. Press the bod firmly into the tap hole, twist the bar to detach the bod, and slide the bar off, turn on the blast and start the melting. The metal will collect in the bottom of cupola with slag floating on top. When sufficient metal is in the well area, the slag will flow from the slag hole. When iron appears at the slag hole, tap the cupola by picking out the bod with a tapping bar: a pointed rod ½ to ¾ inch in diameter with a ring at one end. When enough metal has been tapped, plug the cupola with a fresh bod.

Melting will continue as long as the blast is maintained and metal and coke charges are added. Occasionally, chilled slag will build up around the tuyeres and impede the air blast. They must be cleared by punching them loose with a blunt bar (Fig. 14-10). The cupola can be held in a nonmelting state for several hours. Open the tap hole and drain the well area completely, then shut off the blast and open a tuyere cover. Melting will stop when the blast is shut down. To resume melting, close the tap hole, close the tuyere, and start up the air blast.

The blast pipe is provided with a blast gate to regulate the blast. A blast gate is helpful to keep track of conditions in the cupola. As you may recall, the fuel bed level above the tuyeres is a function of blast pressure: low blast pressure for a low bed, high pressure for a high bed.

To achieve proper melting, the metal must be melted at maximum heat in the fuel bed melting zone, and the oxygen entering the bed from the tuyeres must be entirely converted into carbon dioxide before coming into contact with the metal. If the bed is too low and the metal melts too close to the tuyeres, it will oxidize badly; if the melt takes place too high, it will be cold.

The two ways of telling if the melting is being accomplished correctly is by noting the time it takes for the first metal charge to melt, and by observing the color of the slag; if an apple green is observed upon fracturing the slag, things are going great and you can be assured that the melting is being accomplished in an area devoid of free oxygen. If the slag is dark brown, it contains some oxides of metal; if black, the metal is melting much too low.

What can be done to correct an oxidizing condition? Let's first examine what's happening. The metal is oxidizing because there is free oxygen in the melting zone. This is caused by incoming oxygen not having time to be converted completely to carbon dioxide before reaching the melting metal. This can be due to too low a fuel bed, too high a blast pressure, or both. To correct this condition, first lower the blast pressure, put in an extra 8 inch charge of coke, then add a normal metal charge, followed by 4 inches of coke when the 8 inch split (as it is called) reaches the bed. This can be determined by comparing the amount of metal tapped to the size of the metal charge. If the bed is too high, reduce the amount of coke between, say, two charges of metal; when this coke reaches the bed, it will be lowered by this amount, or raise the blast pressure causing the melting zone to move up. When tapping the metal, watch for excessive sparks (oxidized metal) shooting from the stream issuing from the tap hole; this indicates a low bed. The droplets of iron passing by the tuyere peephole should be bright (hot) without giving off the sparks. If they are dull (cold) the bed is too high, the blast too low, or both; if the droplets are bright, but shoot off sparks, the bed is too low, the blast too high, or both.

After you have done some actual cupola melting and trained your eye to recognize these symptoms, it will all become quite

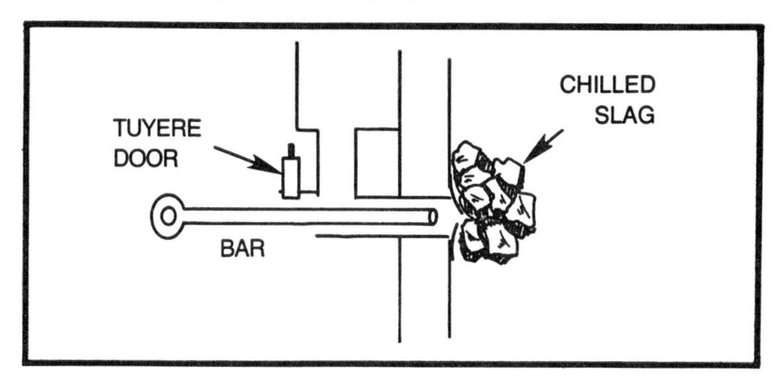

Fig. 14-10. Clearing out the slag.

simple. If there is a foundry in your community that runs a cupola, try to get permission to look it over; watch it operate and talk to the cupola tender. It could be a valuable experience.

When melting bronze, use a low blast pressure (2 ounces average, 3 ounces maximum) and a metal-to-coke ratio of 20 to 1. Tapping temperature should be approximately 2300 degrees Fahrenheit, but no more. If the drops of bronze going past the tuyere peepholes are bright and clear, you are in good shape. If dull, they have become oxidized because of a low fuel bed, excessive blast, or both. If they look cold but don't exhibit signs of oxidation, the bed is too high, the blast too low or both.

When you have reached the end of your melting schedule, tap out all the liquid metal, shut down the blast, open a tuyere hole, pull the bottom door prop out, and allow the burning coke and whatever else is left in the cupola to drop out. Quench the residue with water. When the cupola is cold, you can tell just where the melting was being done by observing the groove burned completely around the refractory lining. The lining must be chipped clean of slag and patched before the next heat. Save the unburned, quenched coke from the drop, as well as any unmelted metal, for the next heat. Do not use drop coke in the next bed of coke; use it between metal charges.

Another common use of the small cupola is making ingots for casting. They can be remelted in the cupola, or in a crucible furnace. One operation I know of has beat the high cost of ingot metal and is making a profit with excess ingots.

They purchase briquets of auto radiators, scrap copper, and red brass scrap from junkyards and melt it in a 16 inch homemade cupola. A typical charge consists of 22 percent red brass scrap, 13 percent scrap copper, and 65 percent essence of radiator. The melt goes on all day, and the resulting metal is made into 17-pound "pigs." They produce semired brass ingots of good color, which they use to do their own casting work. The excess metal is sold for as much as one dollar per pound to other foundrymen, this pays for their freight costs, and provides a ready supply of metal for their area.

Cast iron derived its name from the fact that it is easily cast. It is basically iron containing so much carbon that it is not malleable at any temperature.

A good average tensile strength for plain cast iron is 22,000 pounds per square inch, (psi), and 100,000 psi compressive strength.

The weakening element in cast iron is graphite, the lower the graphite (carbon) in the iron the higher the strength.

To obtain high strength, steel scrap is melted along with the pig iron. The amount of steel in the mix varies from 15 to 60 percent of the charge. The resulting product is called semi-steel, with tensile strengths of 40,000 psi, and still a castable and machineable iron.

The graphite in cast iron is from 11 to 17 percent of its volume. The control of its distribution and flake size is basically controlled by the silicon content and the cooling rate.

The carbon in cast iron is in two forms combined carbon (in solution) and free carbon graphite (not in solution). The free carbon (graphite flakes) is what gives the fractured iron its gray appearance.

As a foundryman wishes to come up with a casting of the desired properties for its end use such as strength, hardness, softness, machineability etc., the basic thing he must control is the proportioning of the combined and free carbon in the casting. As all the carbon is in simple solution when the iron is molten, we must control the amount of free carbon thrown out during solidification to obtain the end results we desire. In proportioning the combined and free carbon in the desired ratio, the three most important factors we must control and consider are the percent of silicon, cooling rate (section thickness) and pouring temperature.

As the percent of silicon increases, the greater the amount of carbon is thrown out as graphite (free carbon) as the metal sets progressively inward from the comparatively cold mold surfaces. The amount of free carbon thrown out is also directly proportioned to the cooling rate. The longer it takes the casting to cool the more free carbon will be thrown out, therefore a thick section will require less silicon and a thin section which chills or cools faster will require more silicon to accomplish the desired proportions. As combined carbon is hard and free carbon (graphite) is soft, if all the carbon stays in combination and no free carbon is thrown out you have white iron: Hard as the hubes of Hades. Castings are often times poured against a chill to cool the outer surface or rim that the face is almost all combined carbon (hard) for wear. Box car wheels have a chilled rim or tire with a soft web and hub, the rim is cast against a chill.

A big problem arises when you have a thin section adjoining a thick section. You can wind up with the thin section as chilled iron and the heavy section soft. This is due to the different cooling rates between the two sections, a poor design.

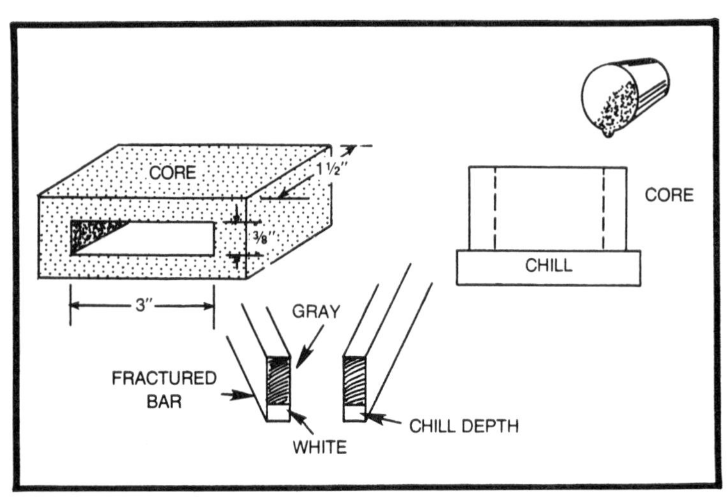

Fig. 14-11. Chill test.

Pouring Temperature: This is the least important, the hotter the pouring temperature the higher the combined carbon will be.

In order for the foundryman to come up with the desired analysis for the intended work he must make up his cupola charge (allowing for losses) by percentages.

Now the foundryman has found from previous experience with his cupola, coke and iron that he will lose 20 percent of the silicon, 10 percent of the manganese and increase the sulphur by 0.03 percent.

Knowing the average analysis of his domestic scrap (sprues, risers, bad castings etc., shop scrap) his purchased scrap (motor blocks, machinery, iron etc.) and the purchased pig iron (sold in a large number of grades from high silicon to low silicon), the foundryman can make the grade of iron he desires. The charge is worked up by multiplying the weight of each kind of material by the percent of the element in it, then divide the total weight of each element by the total weight of the materials that make up the charge.

By the relative adjustments of the pig iron and scrap, mixtures of any desired analysis can be made.

Now you can add additional silicon to the ladle or directly into the stream of metal coming from the cupola to the ladle, to make minor silicon adjustments to control the chill (proportioning of the free and combined carbon). The amount added is established by practice or the use of a chill test. The most common type is a 1½ inch thick core with an open ended cavity ⅜ of an inch thick by 3 inches wide. The core is set on a cast iron plate (chill) and a sample of iron is poured into the chill mold. The cooled casting is fractured, showing the depth of the chill which will appear on the side adjacent to the chill plate. The depth of chill for a given grade of iron is determined by experimentation and experience. More or less silicon can be added at the spout or in the ladle based on the interpretation of the chill test. See Fig. 14-11.

Another type of chill test is the wedge chill test, no chill block used, only a core mold. See Fig. 14-12.

The main thing the unchilled wedge test will tell you is the sensitivity of the metal mix toward chilling. Some melters use both.

The silicon is added in the form of a purchased innoculant of iron silicon. It comes in drums in a granular form.

We have covered mixes etc., and the effect of silicon, and its influence and effect on the carbon, and its ability to cause the iron give up some of its carbon as free carbon (graphite). This is so up to 3 to 3.5 percent,however above this point it acts as a hardener and these irons are known as silvery irons. They contain from 1 to 2 percent carbon and are very brittle. Silvery pig is used to make up mixes with the desired percentage of silicon when the scrap on hand is too low in silicon such as steel. This in granular form is the ferro silicon you use at the spout or ladle for adjustment.

The range of carbon for all practical purposes in pig iron is from 3.25 to 4.25 percent.

CUPOLA CHARGES

Some typical charges for various castings are given below. The symbols used to denote the metals are;

Si..Silicon
S..Sulphur
Mn..Manganese
P..Phosphor
Tc..Total carbon

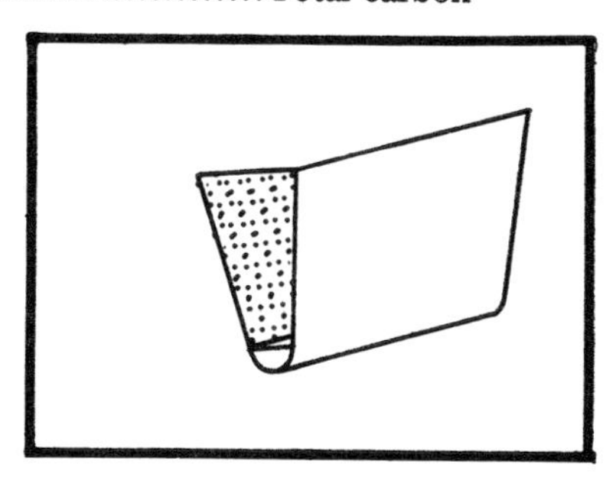

Fig. 14-12. Wedge test.

Product & analysis	**Charge**
Pressure castings	
Tc—3.25%	Pig Iron 50%
Si—1.25%	Steel scrap 25%
Mn—0.65%	Domestic Scrap 25%
P—0.20%	
S—0.10%	
Valves & Fittings	
Tc—3.30%	Pig Iron 50%
Si—2.00%	Steel Scrap 10%
Mn—0.50%	Purchased scrap 40%
P—0.60%	
S—0.10%	
General Castings (Soft)	
Tc—3.40%	Pig Iron 50%
Si—2.60%	Domestic scrap 45%
Mn—0.65%	Steel scrap 5%
P—0.30%	
S—0.10%	
Piston Rings	
Tc—3.50%	Pig iron 60%
Si—2.90%	Scrap or sprues 40%
Mn—0.65%	
P—0.50%	
S—0.06%	
Also a very good mix for light miscellaneous castings.	
Plow Shares (chilled iron)	
Tc—3.60%	Pig iron 45%
Si—1.25%	Steel 15%
Mn—0.55%	Domestic scrap 40%
P—0.40%	
S—0.10%	
The difference in silicon content between piston rings and plow shares.	
Machinery Iron #1 (thin section not over 1 inch thick)	
Tc—3.25%	Pig iron 50%
Si—2.25%	Scrap 50%
Mn—0.50%	
P—0.50%	
S—0.09%	
Machinery Iron #2 (section 1 to 1½ inches thick)	
Tc—3.25%	Pig iron 50%
Si—1.75%	Steel 10%
Ma—0.50%	Scrap 40%
P—0.50%	
S—0.10%	
Machinery Iron #3.25%	Pig iron 50%
Si—1.25%	Steel 25%
Mn—0.50%	Scrap 25%
P—0.50%	
S—0.10%	

You will note that in machinery iron 1, 2, and 3, we called for the same Mn—0.50 and P—0.50, but note the difference in silicon from 1 to 3. Because of the cooling rate difference for the section thickness, silicon promotes free carbon as does slow cooling, thus the adjustment in silicon downward as we get into a slower cooling rate for thick sections.

High Strength Cupola Iron
Tc—2.75
Si—2.25
Mn—0.80
S—0.09

Steel scrap 85%
Returns (sprues, gates, etc.) 10%
Ferro silicon 5%

COMPOSITION OF GRAY IRON CASTINGS

The following is a tabulation of the percentages of additive metals in gray iron castings.

Small Wheels
Si—1.75 to 2.25
S—max 0.08
P—0.40 to 0.60
Mn—0.50 to 0.70

Automobile Castings
Si—1.75 to 2.25
S—max .08
P—0.40 to 0.60
Mn—0.60 to 0.80

Chilled Castings
Si—0.75 to 1.25
S—Max 0.10
P—0.20 to 0.40
Mn—0.80 to 1.20

Agriculture Castings
Si—2 to 2.50
S—max 0.09
P—0.60 to 0.80
Mn—0.50 to 0.80

Permanent Molds
Si—2 to 2.25
S—max 0.07
P—0.20 to 0.40
Mn—0.60 to 0.90

Collars & Couplings
Si—1.75 to 2
S—max 0.08
P—0.40 to 0.50
Mn—0.50 to 0.70

Brake Shoes
Si—1.40
S—max .10
P—0.50
Mn—0.50 to 0.70

Ornamental Iron
Si—2.25 to 2.70
S—max 0.09
P—0.7 to 1
Mn—0.40 to 0.60

Wood Working Machinery
Si—1.75 to 2.25
S—max 0.09
P—0.50 to 0.70
Mn—0.50 to 0.70

Ball Mill Grinding Balls
Si—0.90 to 1.20
S—max 0.15
P—0.20
Mn—0.60 to 1 (Tc low)

Small Pulleys
Si—2.25 to 2.75
S—max 0.08
P—0.60 to 0.90
Mn—0.50 to 0.70

Drop Hammer Dies
Si—1.25 to 1.50
S—max 0.07
P—0.20
Mn—0.60 to 0.80

Locks & Hinges
Si—2.50 to 2.70
S—max 0.08
P—0.70 to 1
Mn—0.40 to 0.60

Steam Radiators
Si—2 to 2.25
S—max .08
P—0.40 to 0.60
Mn—0.50 to 0.70

Soil Pipe & Fittings
Si—1.75 to 2.25
S—max 0.10
P—0.50 to 0.80
Mn—0.50 to 0.80

Novelty Work (toys etc.)
Si—2.50 to 3
S—max .08
P—0.8 to 1
Mn—0.40 to 0.60

Chapter 15
Small Foundry Business or Hobby

If you are a fledgling or novice metal caster and wish to cast for your own amazement, the hobby foundry should be your cup of tea. The hobby foundry can be fully or partially self sufficient. You can cast items which you sell as your product, like a sculptor who casts his own work, or you can cast for the fun of it. Occasionally cast something for someone who pays you for this service. If you are a novice don't take on a job, for pay, until you are sure you can produce, and at a profit, even for a friend.

If you have in mind making a full or part time business out of a small foundry, don't push it, cast, re-cast and learn. If it wants to develop into a small paying enterprise, it will do so on its own. In time, as you learn to produce good work for yourself, the news will spread by word of mouth that you can make this and that. When you reach this point, the decision as to whether you want to take on pay jobs is yours, if so, the kind, size, weight, and metal is up to you. There are numerous backyard small foundries which came about by necessity. For instance, if you need a small casting for a product you make for someone, or sell yourself, but the quantity is too small to interest a production shop, or job shop, where do you turn?

You are faced with either changing the design to eliminate the casting or replacing it with something you can get readily and in small quantities. I know of several cases first hand. One man made an item which consisted of a simple zinc sand casting, and a silver plated spring. This assembly, along with a box and instructions was the entire product. His quantities at the start were not enough to interest a foundry, and if the foundry was interested his cost per unit, due to the small quantities, was more than he could stand, against his retail price for the unit. Knowing full well that the zinc castings in our shop would be a loser, we let him talk us into running them in small quantities. We simply felt sorry for him.

We finally talked him into setting up a small area of his plant to make the zinc castings. We trained one of his people on location. Well, this person would crank up the mini one man foundry about two days a week and mold, melt and cast this zinc part. The rest of the week he did other things, plated springs, assembled etc. As it was a single product he had all his problems worked out in a few days. The plant got their castings in the quantities desired at a cost they could live with. When the business, which flourished, reached the point where the quantities were such that they would attract an outside foundry, they simply expanded their own casting operation. In 10 years they went from one man casting two days a week to a foundry crew of 15 permanent mold operators. They started out with no experience in casting whatsoever, with a little help from us.

In a similar case, I have a pattern maker friend who set up his two grandsons to produce a cast aluminum window fan spider in small lots for a man who makes, in limited quantities, window fans. The boys, both in school, run the foundry on week ends. The profit is put aside for college, or for trips etc. Everybody is happy.

A young man recently came to me about casting replacement parts for people who restore old automobiles, hood ornaments, pistons and what have you. He is going to set up a small foundry for this purpose alone. He knows what is wanted and where the market is and what he can get for his work. He could, in time, make for himself a good paying little enterprise.

As this book is not intended for the professional or the person who declares he is going to learn to sand cast even if it takes a week, my advice is plain and simple. Simply start out slowly. You really don't have to know exactly what your ultimate goal is, if any.

STARTING POINT

In order to learn anything new, you must learn the very basics, and the only place to do this is at the beginning. Let's not think of brass, bronze, aluminum or cast iron or the mixes etc. Let's start at the basic business of making a sand mold and pouring it.

The equipment listed would be the minimum required as a starter.

STARTING EQUIPMENT

Get together the following:

1—Molding shovel—purchase
1—Bench rammer—purchase or make
1—Finishing trowel—purchase
1—Bulb sponge—purchase
1—Medium size spoon & slick—purchase
1—Gate cutter—make
Several sprue sticks—make
Several riser sticks—make
3—12 × 12 × 3″ cope and drag flasks—make
5 lbs. dry parting—purchase
500 lbs. natural bonded molding sand—purchase (for non-ferrous light work)
3—Bottom boards for the 12 × 12 flasks—make
1—Molding board for the 12 × 12 flask—make
Several chunks of cast iron for mold weights—junk yard
1—Molding bench—make
1—#4 Molders riddle—purchase
1—Molding bench—make
1—#8 Molders riddle—purchase
1—Strike off—make
1—Vent wire—make
1—Sprinkler can—hardware store
2—Parting bags (free when you buy parting) or use a sock
1—Flat back wood pattern ½ inch thick 3 inches in dia. disc drafted 1 degree and shellacked—make
1—Draw & rapping pin—make
5 lbs. top grade graphite—purchase

OK, now all we need is something to melt with. Let's start with something simple. Let's melt up zinc die castings, which we purchase from the junk yard. You can melt these with a natural gas burner or propane burner in a cast iron pot. See Fig. 15-1.

A propane or gasoline plumbers pot will work but not as good. In place of a sheet metal ring you can use a sheet metal sleeve lined with fire clay and sand or a castable refractory to cut down heat loss and improve your efficiency. You will also need a cast iron ladle with which to pour.

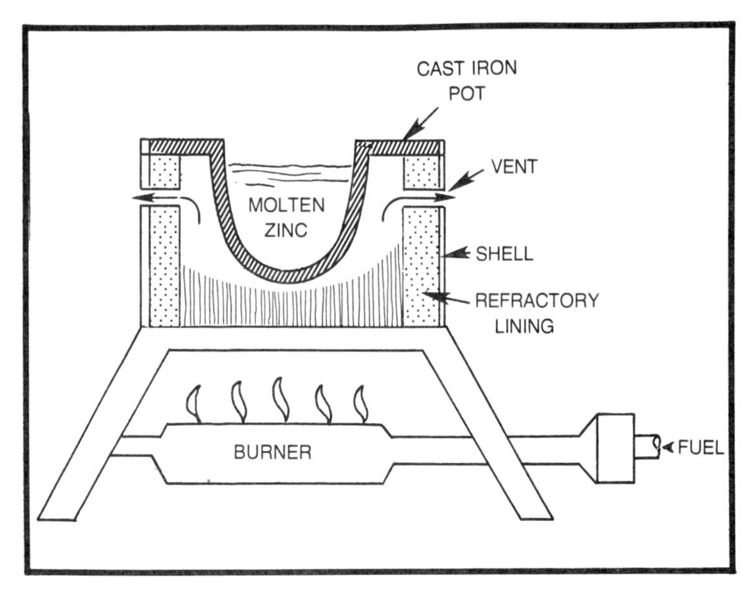

Fig. 15-1. Simple zinc melter.

GETTING STARTED

Now the first day practice conditioning the sand, riddling, cutting, squeeze testing for green strength. Get as familiar with it as possible. To start, sprinkle down 100 pounds, cut, cut, cut, and riddle. If too damp cut in some dry, if too dry sprinkle and re-cut. Work up 200 lbs. or thereabout. When it is in molding condition according to your squeeze test, and it goes through the riddle, not mud but not dust, make a pile by your molding bench. Sprinkle it down and cover with a plastic sheet for the night.

Second Day: You will find the sand feels better after percolating overnight under the plastic. Now place a drag flask, pins down, on a molding board on the bench, (no pattern). Sprinkle with dry parting, riddle, fill, ram and vent, rub in a bottom board and roll it over. Remove molding board and check mold hardness by pressing the sand with your thumb. Go all over looking for soft and hard spots. Repeat this from start to finish until you can not only roll it over easily, but your ramming is nice and even, firm but not hard. You can buy a mold hardness gauge, a hand operated dial indicator which measures the hardness of a sand surface when you press it against the molding sand. And, by moving it around you can see just how hard, soft or uneven you are. But, you are looking at about $85.00. Teach your thumb to do this.

Keep on until you have it down pat. Don't be in a hurry. OK, now ram the drag, roll it over, sprinkle with parting, place a sprue stick in place (no pattern yet). Ram, vent, remove sprue, lift cope, set on set off bench, finish, blow out pouring basin and sprue. Blow off drag and close mold. Oh yes, check cope and drag for hardness and even ramming. OK, do this until you have it down pat. Now try it with your pattern make up the mold completely, close it, then lift off the cope and check for dirt, shake it out and start over. Now, make three molds with your pattern, one gated with one single gate, one gated with a single gate with a side riser opposite the gate, one with a side riser with the sprue gated into riser. See Fig. 15-2.

Melt the zinc and pour one at a time. When cool, shake out and examine them for defects, consult the list in Chapter 11. Saw the gate off and check for shrink.

Now, make up one mold and spray this one with molasses water and skin dry. Make another mold, dust with graphite and slick down with a dry brush and fingers, blow out and close. Make another mold, spray with molasses water and graphite, skin dry. Pour these, examine and compare.

When you have decided on the correct gating, cast three more: one with the mold rammed too hard with no vents brick hard, (*when pouring wear goggles and don't put your face over a sprue or riser*), one rammed too soft and one left dirty. Don't blow out cope or drag just

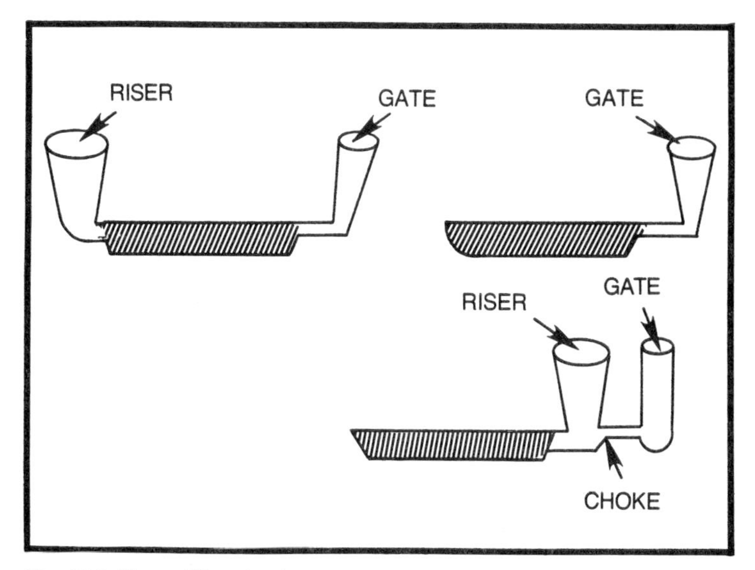

Fig. 15-2. Three different gates.

close and don't riddle the sand on this one. Practice, practice. When you are satisfied that you can cast the disc, move on to a simple pattern that leaves a green sand core, then a dry sand core. As to your foundry, it is a progressive thing, as you learn and get more involved you also add equipment, tools, patterns etc.

When you the reach the point at which you are going to take on some outside work, you will find a lot of price shooters. Why I don't know but some foundries sell castings by the pound. You sell fruit by the pound, not castings. If you tell someone that you charge $3.00 a pound for making red brass castings, you will find that one job weighs 15 pounds in a 12 × 16 inch flask, and the next job is 5 pounds but thin and as long as your arm, and will take a lot more work and time, a larger flask and what have you. You must know exactly your direct and indirect cost for each job. If you buy metal for 60 cents per pound you can't sell the casting for 60 cents per pound, plus labor and overhead. What is your melt loss and degating and grinding loss? 10, 20, 30 percent? Know all your costs and add your profit. If the customer won't pay your price let him go elsewhere.

Above all, do not take on any outside work which does not fit your operation in every way, size, metal quantities, specifications etc., to do so will kill you off quickly.

Every year new small foundries pop up, start off with a bang and go broke. The reasons for their demise are many, the biggest cause is not knowing their actual cost leading to unrealistic pricing. The more you produce, the faster you go broke. Lack of capital and not knowing what you are doing (technology) is last. I know of one case where a man had done some investment casting of jewelry (castings in the ounces and less range). He opened up a good size sand foundry and blew $50,000 up the cupola.

Another aluminum foundry specializing in small match plate work and successful for many years, got greedy and took on a contract (with a performance bond) to produce a mountain of very large complicated close specification aluminum castings. Well when they woke up to the fact that it was not their bag (as they say) they were broke. Their match plate customers were long gone.

Of course this problem is not confined to new and small operations, some biggies who have been at it a long time get crazy and go down the tube for one reason or another. When larger firms go broke, many times it's a case of getting top heavy with non-producers (management experts etc.). The one thing that you cannot sell is salaries, they are not a product.

Expansion will some times do it, get too big with too much fixed overhead to the point that you cannot pull in your horns when you hit a slacker. I've been there. Keep it small and own it all. You don't make money with a foundry, you make castings. Money is the by-product of making good high quality castings.

Glossary of Foundry Terms

absorption—The property of a material to absorb quantities of gases.

aerating—Decreasing the density of the molding sand with the shovel or riddle.

air belt—The chamber around a cupola at the tuyeres to equalize the volume and pressure of the air entering the tuyeres.

air dried—Surface drying of cores in open air prior to baking.

alloy—A mixture of two or more metals.

alloy cast iron—Cast iron alloyed with some other metal such as copper, nickel, cobalt, tungsten etc., to give it the desired characteristics for its intended use.

aluminite—A trade name of the Philips Carey Co. for an aluminum refractory.

arbors—Metal shapes embedded in a green sand or dry sand core for support. A form of reinforcing the core.

artificial sand—Sand that is produced by crushing rock to sand grains.

back draft—Reverse taper on a pattern which prevents its removal from the sand.

backing board—A second board used to rest the cope on its back.

band—A metal band placed inside a snap flask to reinforce the sand at the parting when the flask is removed. The band stays with the mold.

bath—The molten metal in a crucible or furnace.

baume—A measure of specific gravity of liquids. A hydrometer graduated in degrees. Baume is used to measure the specific gravity of liquids.

bayberry wax—The wax extracted from bayberries used to wax patterns.

bed in—Ramming the drag mold, in a flask or pit in such a manner that the drag does not have to be rolled over.

bellows—A hand operated air blowing device.

binder—A material that holds refractory particles together.

black wash—Graphite mixed with water and a binder such as molasses used to coat a core or mold surface (then dried).

blowy—Metal that is full of dissolved gases in the liquid state, and when solid is full of gas pockets.

boss—A projection on a casting of a circular cross section.

brazing—Joining two or more members together with a hard solder composed of copper, zinc and tin.

brittleness—The property of breaking easily under stress.

bronze—A copper-based alloy containing tin, lead and zinc.

bull ladle—A ladle for carrying molten metal, having a shank and two handles.

calcium sulphate—Plaster of paris is hydrated calcium sulphate ($CaSO_4 2H_2O$).

cast (vb)—To pour molten metal into a mold.

casting (noun)—A metal object cast to a required shape rather than machined.

cast gate—The ingate through which the metal is cast.

casting ladle—A crucible for conveying molten metal from a furnace, and for pouring it into a mold.

chemical composition—The percentage of the elements making up a composition or alloy.

chaplets—Metallic nail-like supports or spacers used in a mold to hold cores in their proper position during the casting process.

charge—The metal placed in a furnace to be melted.

chill—A device, usually metal, used to promote the solidification of molten metal to prevent shrinkage at the point applied.

cold shortness—The opposite of hot shortness, usually caused by high phosphorous, dissolved gases and oxides.

combined water—Water chemically combined with mineral matter; it is driven off only above 500°F.

combustibility—The property of a substance to oxidize with the evolution of light and heat.

contraction—The property of a substance to reduce its volume under changes of temperature (negative expansion). Do not confuse this with shrinkage.

core raise—A casting defect caused by the core floating to the top of the mold due to the improper use of chaplets or prints.

corrosion—Slow oxidation or the wasting away of a metal.

critical temperature—The temperature at which a substance is physically changed.

crucible—A ceramic pot made of graphite and clay, or clay and a refractory material, used to contain metal being melted.

crystalline—Any material having a regular molecular structure.

crystobalite—A form of quartz with a melting point of 3140°F; crushed, it is used as a refractory material.

decrepitation—The property of a substance to fly apart on being heated (explode).

density—The mass of a unit volume of a material at different temperatures.

drop—A casting defect, the result of a portion or a projection in the mold breaking and dropping into the mold.

dross—Metal oxides in, or on, molten metal.

ductility—The property of a substance to be permanently deformed without rupture.

ethyl silicate—The principal binder for ceramic investment molds (an ester mixed with silica flour).

ferrous—Containing iron.

fillet—A concave shape at the corners of intersections within a pattern.

firebrick—Brick made of high-refractory clays for linings in furnaces and building ovens or kilns.

fluidity—The property to flow easily when in a liquid state. Melted; cast iron, aluminum, lead, brass etc., are very fluid where as wrought iron cannot be made sufficiently fluid to cast. Oxidized, gassy and improper melting practice reduces the fluidity.

fluxing—Applying a solid or gaseous material to molten metal to remove oxides.

founding—The art of casting metals.

foundry—A shop in which casting metals is the prime endeavor.

foundryman—Craftsman skilled in casting metals.

fused silica—Silica that has been fused at a high temperature and ground into various meshes (particle sizes) for use as a refractory. It is used as a stucco for ceramic shell molds, and for full-mold investment casting.

gagger—A rod used to reinforce or support a portion of a mold.

gate—The hollow end of a runner where metal enters a casting.

gated pattern—A pattern with gates attached.

graphite—Thin plates or flakes of pure carbon.

hardness—The resistance to permanent deformation or abrasion as measured by Brinell, Rockwell or Seleroscope instruments.

hardness (cont'd)—The Brinell and Rockwell machines measure the hardness by measuring the indentation made on the material by a steel ball under a load. Seleroscope measures the hardness by the rebound height of a steel ball dropped from a given height on to the object tested. The information from these tests are called hardness values.

head—Opening in the top of a mold; also, the pressure exerted by molten metal.

horn gate—A gate designed to allow metal to enter the mold at the bottom center of the mold cavity.

hot shortness—The loss of strength at high temperature as compared at ordinary temperatures, easily broken when hot but not when cold. (ordinary red brass is very hot short).

inclusions—Particles of impurities in the casting.

investment compound—A refractory used for molds formed around an expendable pattern.

ladle—A metal receptacle lined with refractories and used for pouring or transporting molten metal.

liquid contraction—Shrinkage occurring in liquid metal as it cools.

lithium—The lightest of all metals and extremely unstable. It combines readily with oxygen, nitrogen, and sulphur, forming low melting compounds which pass off as gases; it is widely used as a deoxidizer of molten metals.

lost wax process—A casting process in which a wax pattern is used. The wax pattern is invested in a refractory slurry; after the mold is dry the wax is melted out, forming a cavity into which metal is poured.

machinability—The relative ease of machining a material.

melting loss—Loss of metal in the charge during melting.

melting point—The temperature at which a solid changes to a liquid.

mold—A form containing a cavity into which metal is poured.

molding sand—Any sand employed for making molds used to cast metals; sand that is a refractory and contains clay as a binder.

mullite—A mineral originally found on the Isle of Mull, it is used as a refractory material for firebricks and furnace linings.

nonferrous—Containing no iron.

olivine—A translucent mineral usually occurring in granular form, used as a refractory in making molds or as a molding sand.

oxidation—Any reaction in which an element combines with oxygen.

permeability—The property of matter which permits gas or liquid to pass through it.

pickling—Cleaning metal with an acid to remove dirt, scale, oxides, etc.

plasticity—The property of sustaining permanent deformation without rupturing.

pouring—The transfer of melted metal from a ladle or crucible into molds.

pouring basin—An opening in a mold into which molten metal is poured.

pyrometer—An instrument used for measuring temperature.

refractory (non)—A heat-resistant material, usually non metalic, used for making furnace linings, molds, and crucibles.

refractory (adj.)—The quality of resisting heat.

riddle—A form of sieve for sifting sand or other particular materials; or, the act of sifting.

rigidity—Stiffness, resisting a change in shape.

riser—A reservoir of excess molten metal at the top of the heaviest section or sections of a casting; it is designed to supply metal that cannot be fed properly from a gate, thus compensating for shrinkage.

shakeout—The operation of removing a casting from a mold.

slicker—Molder's tool used for smoothing a mold surface.

solubility—The chemical property of a substance, gas etc., to dissolve in a liquid.

southern bentonite—A clay used in various ceramic formulas and refractories, and as a binder for foundry molding sands.

specific gravity—The weight of a unit volume, referred to the weight of an equal volume of water at 4° centigrade.

talc—A hydrated magnesium silicate.

water glass—Silicate of soda; it is used as a binder in some molds and cores.

wearing properties—The resistance to frictional forces.

wild metal—Metal that boils in the ladle and will not remain quiet, usually resulting from insufficient or too much deoxidation (treatment) faulty melting.

zircon—Zirconium silicate; it is used as a refractory for molds, firebricks, and crucibles.

Appendix A
Fluxes, Pickles, and Dips

The following is a collection of fluxes, pickles, dips and furnace linings that I have collected over the years, hobo fashion, and all have been foundry tested.

BRIGHT DIP FOR ALUMINUM CASTINGS

Dip in diluted solution of caustic soda, then dip into a solution of 50% nitric acid and 50% water. Wash in clear water and dry in hardwood sawdust.

PICKLES FOR BRASSES

1. Nitric acid.................................1½ parts
   Sulphuric acid.................................2 parts
   Sodium chloride ....2 oz for each 4 gal
2. Sulphuric acid.................................5 parts
   Saltpeter.................................1 part
   Water.................................2.5 parts
3. Yellow nitric acid.................................1 part
   Sulphuric acid.................................1 part
   Sodium chloride.................................½ oz per gal
   Rinse in clean hot water containing a small amount of cream of tartar, dry in hardwood sawdust.

PLASTIC REFRACTORY

Brick grog 70%
Fire clay.................................30%
Mix dry before adding water.

BRASS FURNACE LINING

Wisconsin ganester 3/8 inch mesh 50%
Wisconsin ganester 1/8 inch mesh 30%
Silica flour 14%
Bentonite 6%

CASTABLE REFRACTORY

Grog #3 mesh 70%
Aluminite cement 25%
Fire clay 5%

MOLD SPRAY

Glutrin water to which graphite or talc is added to correct Baume. Good for ferrous and nonferrous.

CORE WASH

11 parts iron oxide, talc, or plumbago
1 part benonite
Molasses water mixed to correct Baume

BRASS FLUXES

1. 100 lb dehydrated borax
   77 lb whiteing
   50 lb sodium sulphate
2. 50% Razorite
   50% soda ash
3. 8 parts flint glass
   1 part calcined borax
   2 parts fine charcoal
4. 5 parts salt
   5 parts sea coal
   15 parts sharp sand
   20 parts bone ash
5. 25% soda ash
   25% plaster of paris
   25% fine charcoal
   25% salt
6. 50 lb glass
   55 lb Razorite
   5 lb lime

CORE PASTE

50% foundary molasses
50% talc

OLD TIMER FLUX FOR COPPER

Mix equal parts of charcoal and zinc with enough molasses water to form a stiff paste. Roll into balls about 2 inches in diameter, then dry. When the copper just starts to melt drop in enough balls to give a good cover.

TOUGH-JOB CORE WASH

4 parts silica flour
1 part delta core wash (50° Baume for dipping and 40° for spraying)

BRAZING FLUX

50% boric acid
50% sodium carbonate

GERMAN SILVER FLUX

5 parts ground marble
3 parts sharp sand
1 part borax
1 part salt
10 parts fine charcoal

BRIGHT DIP FOR COPPER CASTINGS

10 gal water
1 gal sulphuric acid
10 lb potassium dichromate

Wash after bright dip and neutralize acid by dipping in a 1% sodium carbonate solution, then a final rinse.

Appendix B

General Mixes

Aluminum Bronze
Cu—90%
Al—10%

Brazing Metal
Cu—84-86%
Zn—14-16%
Fe—0.06% Max.
Pb—0.30%

P.S. (Plumber's Special)
Cu—80%
Pb— 6%
Sn— 3%
Zn—11% semired bronze

Grille Metal
Cu—69%
Sn— 1%
Zn—30%

Journal Bronze
Cu—82-84%
Sn—12.5-14.5%
Zn—2.5-4.5%
Fe—0.06 max
Pb—1 max

Imitation Manganese Bronze
Cu—59 lb
Zn—40 lb
Al—5 oz
Sn—5 oz
Tin plate clippings—6 oz.

Cheap Mix for Plumber's Ware
Sheel metal clippings—71½%
Copper wire—25%
Pb—2%
Sn—1%
P Cu (15%) 8 oz per
100 lb of above.

Cheap Red Brass
Scrap copper wire—40 lb
Zinc—7.5 lb
Lead—7.5 lb
Mixed brass scrap—45 lb
Melt mixed brass scrap add
copper, zinc, and lead.

Another Red Brass for Small Castings
Cu—82% Zn—8%
Pb— 8% Sn—2%
Not used for steam work

Munts (General Spec)
Cu—59-62%
Zn—39-41%
Pb—0.6 max

Hard Bearing Bronze
Cu—63.5%
Zn—21.5%
Manganese—4%
Fe—4%
Al—7%

Commercial Yellow Bronze
Cu—69%
Pb— 3%
Sn—1.5%
Zn—26.5%

Statuary Mix #1
Cu—90%
Sn— 7%
Zn— 3%
8 oz of Pb per 100 lb

Statuary Mix #2
Cu—90%
Sn— 5%
Zn— 5%
8 oz of Pb per 100 lb

Statuary Mix #3
Cu—90%
Zn—7.5%
Sn—2.5%
Also used as a tablet alloy

Statuary Mix #4
Cu—88%
Sn— 6%
Zn—3.5%
Pb—2.5%

Commercial Brass
Cu—64-68%
Zn—32-34%
Fe—2% max
Pb—3% max

Gun Bronze
Cu—87-89%
Sn—9-11%
Zn—1-3%
Fe—0.06% max
Pb—0.30% max

Ornamental Bronze
Cu—83%
Pb— 4%
Sn— 2%
Zn—11%

Red Ingot
Cu—85%
Pb— 5%
Sn— 5%
Zn— 5%

Pressure Metal
Cu—83%
Pb— 7%
Sn— 7%
Zn— 3%

Gear Bronze
Cu—87.5%
Pb—1.5%
Sn—9.5%
Zn—1.5%

Bronze
Cu—89.75%
Sn—10%
Ph—0.25%

88-4 Bronze Ingot
Cu—88%
Sn— 8%
Zn— 4%

Bearing Bronze
Cu—80%
Pb—10%
Zn—10%

Lead Lube
Cu—70%
Pb—25%
Sn— 5%

Manganese Bronze
Cu—60%
Zn—42%
Balance temper depending on grade desired

Nickel Silver
Cu—61%
Zn—20%
Ni—18%
Fe— 1%

Ship Bells
Cu—82%
Sn—12%
Zn— 6%

Cheap Yellow Brass
Cu—60%
Zn—36%
Pb— 4%

Another Cheap Red Brass
Cu—83%
Pb—8.5%
Zn—6.5%
Sn— 2%

Yellow Brass for Small Castings
Cu—70% Sn—2.8%
Zn—25.5% Pb—2%

Half & Half Yellow—Red Brass
Cu—55%
Zn—10%
Pb— 5%
Yellow brass scrap-30%

Tough Free Bending Metal
Cu—84.5%
Zn—10%
Pb— 3%
Sn—2.5%

Appendix C

Bell Mixes

(WHITE)TABLE BELLS

Sn—97%
Cu—2.5%
Bi—0.5%

SWISS CLOCK BELLS

Cu—75%
Sn—25%

SILVER BELLS

Cu—40%
Sn—60%

BEST TONE

Cu—78%
Sn—22%

HOUSE BELLS

Cu—78%
Sn—20%
Yellow Brass—2%

SLEIGH BELLS

Cu—40%
Sn—60%

HIGH GRADE TABLE BELL

Sn—19%
Ni—80%
Pt— 1%

SPECIAL CLOCK BELLS

Cu—80%
Sn—20%

SPEC. SILVER BELLS

Cu—50%
Zn—25%
Ni—25%

GENERAL BELL

Cu—80%
Sn—20%

FIRE ENGINE BELLS

Sn—20%
Ni— 2%
Si— 1%
Balance Cu

LARGE BELLS

Cu—76%
Sn—24%

RAILROAD SIGNAL BELLS

Cu—60%
Zn—36%
Fe— 4%

HOUSE BELLS

Cu—76%
Sn—16%
Yellow Brass—8%

GONGS

Cu—82%
Sn—18%

Appendix D
Miscellaneous Metal Mixes

SLUSH METAL

Zinc....................92%
Aluminum....................5%
Copper....................3%

ORIENTAL BRONZE

Copper....................84%
Lead....................10%
Tin....................5%
Zinc....................1%

IMITATION GOLD

Copper....................89%
Zinc....................10.5%
Aluminum....................0.5%

ALUMINUM BRONZE

Copper....................90%
Iron....................05%
Aluminum....................9%
Manganese Copper....................05%

PHOSPHOR BRONZE

#1 Copper Wire..90%
Phosphor Tin...10%

IMITATION SILVER

Copper...50%
Nickel...20%
Zinc..30%

HIGH PRESSURE BRONZE

Copper...84%
Tin...7.5%
Lead...5.5%
Zinc..3%

CASKET METAL TRIM

Tin...0.5%
Antimony 12.5%
Arsenic 8.5%
Lead 87.5%

ALUMINUM SOLDER

Tin...68%
Antimony..0.5%
Zinc 21.5%

Appendix E
Approximate Analysis
Various Grades Scrap Aluminum

Material	Cu	Fe	Zn	Si	Misc
Crankcase	7–8	1–1.25	1–1.5	1–1.5	
Misc Cast	7–8	1–1.25	1–1.5	1–2	
Cable	0–0.10	.2–.50	0–0.2	0.2–0.4	
Pistons	8–9	0.76–1.25	0.1–0.5	0.75–1.25	
New Clips	0.1–0.3	0.3–0.8	0–0.2	0.2–0.7	
Alum Die Cast	6–7	1–1.5	1–2	2–3	Ni 0.5–1.0
Dural Clips	4–4.5	0.5–1.	0–0.25	0.25–0.5	Mg 0.5–1.0
					Mn 0.5–1.5
Pots—Pans					
Dishes	0.25–0.50	0.50–0.60	0.25–0.50	0.25–0.50	

Appendix F
Judging the Approximate Temperature of an Object in the Dark, by Eye:

Judging the approximate temperature of an object in the dark, by eye:

Faint Red	878°F
Dull Blood Red	990°F
Full Blood Red	1050°F
Dull Cherry Red	1195°F
Full Cherry	1375°F
Light Cherry	1550°F
Deep Orange	1640°F
Light Orange	1730°F
Yellow	1832°F
Light Yellow	1976°F
White	2200°F
Bright White	2550°F
Dazzling White	2730°F

Appendix G
Sources and Supplies

FOUNDRY TOOLS AND EQUIPMENT

M. A. Bell Co.
217 Lombard St.
St. Louis, Mo.
(Branches in Major Cities)

Hill & Griffith Co.
Birmingham, Ala.
(Branches in Major Cities)

Firebrick and Refractories:
A.P. Green Co.
Mexico, Mo.
(Branches in Major Cities)

WAX

Freeman Supply Co.
1152 East Broadway
Toledo, Ohio 43605
Branches in Major Cities
(Wax, sprue wax, pattern supplies & equipment)

Wax Company of America
5016 West Jefferson Blvd.
Los Angeles, Calif. 90016

ALBANY & FRENCH SANDS

New York Sand and Facing Co.
106-114 Grand Ave.
Brooklyn New York

Albany Sand
French Sand

PREPARED INVESTMENTS

C. W. Ammen & Son
966 Manitou Ave.
Manitou Spgs., Colo. 80829
"Ammen's Super Investment Compound"

Ransom & Randolph Co.
Toledo, Ohio 43601

Kerr Mfg. Co.
1111 E. Milwaukee
Detroit, Mich.

BRONZE FURNACES

Complete bronze furnaces for studio foundries and miscellaneous equipment.

C. W. Ammen & Son
966 Manitou Ave.
Manitou Springs, Colo. 80829

KILNS

Denver Fire Clay Co.
Denver, Colo.

Kilns Supply & Service Corp.
P.O. Box Q
Mamaroneck, N.Y. 10543

SCULPTOR SUPPLIES

Wax tools, clay tools, and just about anything needed by the sculptor.

Sculpture House
38 E. 30th St.
New York, N.Y. 10016

BRONZE & BRASS

American Smelting & Refining Co.
120 Broadway
New York, N.Y.

RUBBER

Chicago Latex
3017 West Montrose Ave.
Chicago, Ill. 60618

Permaflex Mold Co.
1919 E. Livingston Ave.
Columbus, Ohio 43209

INDEX